Statistical Fluid Dynamics

Statistical Fluid Dynamics

Editors

Amine Ammar
Francisco Chinesta
Rudy Valette

MDPI • Basel • Beijing • Wuhan • Barcelona • Belgrade • Manchester • Tokyo • Cluj • Tianjin

Editors
Amine Ammar
HESAM Université
France

Francisco Chinesta
HESAM Université
France

Rudy Valette
PSL Research University
France

Editorial Office
MDPI
St. Alban-Anlage 66
4052 Basel, Switzerland

This is a reprint of articles from the Special Issue published online in the open access journal *Entropy* (ISSN 1099-4300) (available at: https://www.mdpi.com/journal/entropy/special_issues/Fluid_Dynamic).

For citation purposes, cite each article independently as indicated on the article page online and as indicated below:

LastName, A.A.; LastName, B.B.; LastName, C.C. Article Title. *Journal Name* **Year**, *Volume Number*, Page Range.

ISBN 978-3-0365-4655-1 (Hbk)
ISBN 978-3-0365-4656-8 (PDF)

© 2022 by the authors. Articles in this book are Open Access and distributed under the Creative Commons Attribution (CC BY) license, which allows users to download, copy and build upon published articles, as long as the author and publisher are properly credited, which ensures maximum dissemination and a wider impact of our publications.

The book as a whole is distributed by MDPI under the terms and conditions of the Creative Commons license CC BY-NC-ND.

Contents

About the Editors vii

Amine Ammar, Francisco Chinesta and Rudy Valette
Introduction for the Special Issue: Statistical Fluid Dynamics
Reprinted from: *Entropy* **2022**, *24*, 782, doi:10.3390/e24060782 1

Abderrahim Bakak, Mohamed Lotfi, Rodolphe Heyd, Amine Ammar and Abdelaziz Koumina
Viscosity and Rheological Properties of Graphene Nanopowders Nanofluids
Reprinted from: *Entropy* **2021**, *23*, 979, doi:10.3390/e23080979 7

Ali Daher, Amine Ammar, Abbas Hijazi and Lazhar Benyahia
Effect of Shear Flow on Nanoparticles Migration near Liquid Interfaces
Reprinted from: *Entropy* **2021**, *23*, 1143, doi:10.3390/e23091143 25

Nesrine Aissa, Louis Douteau, Emmanuelle Abisset-Chavanne, Hugues Digonnet, Patrice Laure and Luisa Silva
Octree Optimized Micrometric FibrousMicrostructure Generation for DomainReconstruction and Flow Simulation
Reprinted from: *Entropy* **2021**, *23*, 1156, doi:10.3390/e23091156 45

Toufik Boubehziz, Carlos Quesada-Granja, Claire Dupont, Pierre Villon, Florian De Vuyst and Anne-Virginie Salsac
A Data-Driven Space-Time-Parameter Reduced-Order Model with Manifold Learning for Coupled Problems: Application to Deformable Capsules Flowing in Microchannels
Reprinted from: *Entropy* **2021**, *23*, 1193, doi:10.3390/e23091193 61

Rabih Mezher, Jack Arayro, Nicolas Hascoet and Francisco Chinesta
Study of Concentrated Short Fiber Suspensions in Flows, Using Topological Data Analysis
Reprinted from: *Entropy* **2021**, *23*, 1229, doi:10.3390/e23091229 87

Claire Dubot, Cyrille Allery, Vincent Melot, Claudine Béghein, Mourad Oulghelou and Clément Bonneau
Numerical Prediction of Two-Phase Flow through a Tube Bundle Based on Reduced-Order Model and a Void Fraction Correlation
Reprinted from: *Entropy* **2021**, *23*, 1355, doi:10.3390/e23101355 103

Xin Ma, Jianmin Zhang and Yaan Hu
Analysis of Energy Dissipation of Interval-Pooled Stepped Spillways
Reprinted from: *Entropy* **2022**, *24*, 85, doi:10.3390/e24010085 125

Hua-Shu Dou
No Existence and Smoothness of Solution of the Navier-Stokes Equation
Reprinted from: *Entropy* **2022**, *24*, 339, doi:10.3390/e24030339 139

Ruifang Shi, Jianzhong Lin and Hailin Yang
Numerical Study on the Coagulation and Breakage of Nanoparticles in the Two-Phase Flow around Cylinders
Reprinted from: *Entropy* **2022**, *24*, 526, doi:10.3390/e24040526 151

About the Editors

Amine Ammar

Amine Ammar is a full professor of Computational Mechanics at Arts & Métiers Paris Tech (French High School of Engineering). He obtained his PhD Thesis in 2001 at Paris VI university—ENS Cachan; the title is "Polymer flow induced crystallization, Application to injection molding". In 2006, he obtained his French degree to supervise research (HdR), specializing in mechanics. His focus was on "the solution of kinetic theory models related to complex fluids". In 2009, he received the scientific prize from the European Association of Material Forming, awarded during the 12th Esaform conference, and the French Jean Mandel prize, awarded during the 19th CFM conference. His research has focused on different topics, such as model reduction in PDE resolution, proper generalized decomposition, kinetic theory of polymers and suspensions, and short reinforced fiber composite processing. He is the author of more than 100 scientific works, published in international journals. He has supervised 20 PhD theses. He is the assistant director of the research laboratory LAMPA—Angers, and the local correspondent at Angers of the doctoral school of Arts & Métiers.

Francisco Chinesta

Francisco Chinesta is currently a full professor of computational physics at the ENSAM Institute of Technology (Paris, France), Honorary Fellow of the "Institut Universitaire de France"—IUF- and Fellow of the Spanish Royal Academy of Engineering. He is the president of the ESI Group scientific committee and director of its scientific department. He was (2008–2012) AIRBUS group chair professor and, since 2013, he has been the ESI group chair professor on the advanced modeling and simulation of materials, structures, processes and systems. He has received many scientific awards (including the IACM Fellow award, the IACM Zienkiewicz award (New York, 2018), the ESAFORM award, etc.). He is the author of more than 350 papers in peer-reviewed international journals and more than 1000 contributions to conferences. He was president of the French association of computational mechanics (CSMA) for 8 years, and chair of the WCCM in Paris (2020). He is director of the CNRS research group (GdR) on model-order reduction techniques in engineering sciences, and the editor and associate editor of many journals. He has received many distinctions, including the Academic Palms, and the French Order of Merit. In 2018, he received Doctorate Honoris Causa at the University of Zaragoza (Spain) and, in 2019, the silver medal from the French CNRS. He is the director of the DESCARTES project on Hybrid Artificial Intelligence that the CNRS is developing in its hub at Singapore (35 M€, 5 years and more than 160 researchers).

Rudy Valette

Rudy Valette is a full professor of Fluid Mechanics and Rheology at MINES-ParisTech, PSL Research University (French High School of Engineering). He obtained his PhD Thesis in 2001 at MINES-ParisTech; the title is "Stability of the multilayer Poiseuille flow of viscoelastic fluids, application to the coextrusion process". He received a PhD prize from the French Society of Rheology. In 2014, he obtained his French degree to supervise research (HdR), specializing in physics. His degree focused on the "Rheology of complex fluids and application to mixing and extrusion processes". His research is devoted to modelling non-Newtonian fluid flows, using experimental and computational methods.

Editorial

Introduction for the Special Issue: Statistical Fluid Dynamics

Amine Ammar [1,*], Francisco Chinesta [2] and Rudy Valette [3]

1. ESI Chair @ LAMPA, LAMPA, Arts et Metiers Institute of Technology, HESAM Université, CEDEX 01, F-49035 Angers, France
2. ESI Chair @ PIMM, PIMM, CNRS, CNAM, Arts et Metiers Institute of Technology, HESAM Université, F-75013 Paris, France; francisco.chinesta@ensam.eu
3. MINES-ParisTech, CEMEF, PSL Research University, F-06904 Sophia Antipolis, France; rudy.valette@mines-paristech.fr
* Correspondence: amine.ammar@ensam.eu

Characterizing complex material consists in establishing the relationship between flow rheology during forming processes and the induced micro-structural state that affects directly the final mechanical properties of the formed parts. It is necessary that research activities reach to address the coupling between forming process and mechanical performances (e.g., fatigue or reliability). Even if in this issue we put the main attention on the fluid dynamic part, the research activity today must cover the life cycle from individual kinematics of particles to the mechanical properties of formed parts. Some points of illustration are quoted here on a non-exhaustive basis:

- Individual kinematics of one fiber or particle in non-Newtonian flow and consideration of hydrodynamic interactions remains today an open subject experimentally as well as numerically (using direct approaches). No accurate analytical characterizations have been done in non-Newtonian fluids in infinite domains. A specific attention must be considered for viscoelastic materials. Such a material is usually encountered during forming processes of injection molding or thermoplastics extrusion.
- After this first step, one must be able to make statistical transitions in order to predict the orientation distribution of a suspension of particles. This step suppose that the individual kinematics has been well established and must take into account particles interactions in order to predict different phenomena such as aggregation.
- Composites involve in their liquid or solid state a significant fluctuation of physical properties due to their heterogeneity. In order to get a fine description, numerical models use generally an excessive computing time and requires a high capacity of storage. This is a consequence of the high number of degrees of freedom requested to correctly describe the physical properties. It will then be necessary to use some approaches that reduce the numerical cost, such as the POD (Proper Orthogonal Decomposition) or the PGD (Proper Generalized Decomposition). These methods will make simulations of complex flows easier. The main difficulty consists in predicting correctly the microscopic state that affects directly the final properties. Two modeling scale are to take into account: the first one is related to the global dimension of the flow providing velocity field, thermal distribution, pressure and other macroscopic fields. The second one is related to the orientation of fibers, the conformation of molecules, etc.
- Establishing distribution models is an essential information for reliability models. Reliability of composites have to be studied through the part life cycle. This requires reproducing (numerically and experimentally) the succession of loadings inducing material damage. Modeling aspects have to be oriented to establish the tools allowing to predict failure probability. These tools could be built by using the statistical information previously calculated.

- Next, fatigue and damage models using internal variables have to be related to the induced microstructure. Indeed, modeling of fatigue is constraint by the CPU time and requires to simulate a very high number of cycles. Two strategies are possible in order to circumvent the difficulty related to the computational cost of the temporal scale: (i) the first one consists in making a decomposition of the time dimension into two dimensions, where the first one is related to a finite small number of cycles and the other one is related to a global evolution of the internal variables. (ii) And the second strategy consists to come back to POD-like techniques which are suitable for extracting modes in cyclic behavior.
- These approaches must be associated to homogenization procedures for complex materials. A specific knowledge of space scales transitions and the relationship with the Representative Elementary Volume (REV) during the forming process or the mechanical loadings, is an essential information for using composites material during their life cycle. Fatigue models in direct simulations could be compared to fatigue model with homogenized variables.
- Machinability of complex materials is also an interesting subject. Modeling the cutting process and the confrontation with experimental measures could give an idea to bring a multi-physic comprehension of chip formation and the tool/workpiece interaction by adopting finite element approaches and methodologies. Microscopic state is determinant in these conditions.

In all this sequence just described the challenge in the flow phase consists to establish the relationship between the flow rheology during the flowing process and the state of the induced micro-structure that directly affects the quality of the obtained mechanical parts. This characterization should take into account the multi-scale description of the continuum matter. Many developments are required. One of them consists in identifying experimentally, numerically, and analytically the laws that govern each scale rigorously. Another challenge consists in developing numerical techniques that allow addressing a detailed description of the physical laws involving a large number of degrees of freedom. Development and control of advanced numerical techniques and experimental observation is essential to predict accurately and with a lowest cost the state of matter and the resulting properties.

Achieving the goal of modeling micrometric and nanometric suspensions remains a major issue. This help to master in a controlled way the mechanical, thermal, and electrical properties, among others, of the suspensions and then of the resulting product when considered in material forming, flow of heat transfer fluids or other applications. In some cases, they can contribute to improve the performance of energy transport. An optimal use of these products is based on an accurate prediction of the flow-induced properties of the suspensions and consequently of the resulting products and parts.

The scientific issues to solve are mainly related to the prediction of the behavior evolution. Particles suspended in a viscous medium tend to modify the behavior. The final properties of the resulting microstructured fluid or solid become radically different from the simple mixing rule. There are numerous works addressing homogenization strategies for systems consisting of perfectly of dispersed particles in a matrix. However, in most cases, particles aggregate or sediment, or exhibit strong induced anisotropy. The microscopic description, despite being the finest one, is too heavy from both computational and experimental points of view. For this reason, coarser descriptions are sometimes preferred. Even if they are less accurate, they lead to faster simulations.

Considering the general behavior of suspensions, viscoelastic fluids or complex flows, two levels of description are relevant: a level related to the overall kinematics and a level associated with the material point in the microscopic scale [1].

Taking into account the state of microscopic structure can be done at different scales. For some behaviors such as fibers, differential approximation requires a closure relationship [2]. Unfortunately, in most cases (except for some special ones), there is no equivalence between the constitutive laws and the microscopic definition of the structure. A microscopic simulation at the scale of the kinetic theory is then required.

The most common technique for kinetic theory problems is the stochastic approach. A lot of work has been done on different models of kinetic theory (dumbbell models, fibers, polymer melts ...) see, for example [3,4].

For higher dimensions (much higher than three) stochastic methods become limited. In the few studies that we find in the literature on the simulation of this type of problem, the authors use a discrete approach (Brownian or with Monte Carlo) that involves the use of a large number of particles. In very special cases, the probability distribution evolution can be expressed as N evolution problems of N different functions with a vectorial change of variable.

The framework for these problems requires the development of specific numerical techniques applicable for problems with large numbers of degrees of freedom.

The difficulty of multidimensional models resolution is related to the proposal of new appropriate strategies able to circumvent the curse of dimensionality. One possibility lies in the use of sparse grids [5]. However, as argued in ref. [6], the use of sparse grid is restricted to models with moderate multidimensionality (up to 20). Another technique able to circumvent, or at least alleviate, the curse of dimensionality consists in using a separated representation of unknown fields (see ref. [7]. for some numerical elements on this topic).

The question of multidimensionality has also been subject of works related to the space-time separated representation. In fact, such decompositions were proposed many years ago by Pierre Ladeveze as an ingredient of the powerful non-linear-non-incremental LATIN solver that he proposed in the 80s. During the last twenty years many works were successfully accomplished by the Ladeveze group. The interested reader can refer to [8] and the references therein related to the radial approximation, denotation given to the space-time decomposition in the LATIN framework.

The resolution of problems with analytical solution is rarely possible. Analytical solution is provided only for specific simplified equations. Otherwise, solution is searched for as a discrete form over a given set of points. Once the discrete solution is obtained in these points the continuum solution can be built on using an appropriated interpolation. When model is defined in dimension D, and with N degrees of freedom in each direction the resolution requests N^D discrete points. The difficulty related to the information processing and storage becomes exponentially dependent on the dimension D. Beyond the value of D equals to 3, standard discretization techniques (such as finite elements, finite differences, or finite volumes methods) suffer from the limitations related to the high number of degrees of freedom.

On this numerical point of view, some contributions have been focused on the development of a new strategy different from the classical based-mesh techniques (FEM, FDM, FVM). The developed method called the PGD (Proper Generalized Decomposition) allows circumventing the curse of dimensionality and allows particularly to solve space-time problem avoiding the use of standard incremental time scheme. It allows more generally to solve problems defined in multidimensional space. The main idea consists to build up the multidimensional solution as a tensor product of functions expressed in lower dimensions.

This strategy was useful for transient problem, but also has proved its robustness for solving micro-macro problems by including configuration and physical spaces in the same discretization. In addition, this strategy has been successfully applied for solving kinetic theory problem when configuration space dimension exceeds the value three. This situation in encountered in the kinetic theory of melt polymer or for bead-spring-chain model for polymer suspension.

The obtained results as well as the potential application are encouraging to carry on the development of this technique and to improve its performance in terms of convergence speed and optimality and to enlarge its application fields.

In addition, considering in general the behavior of statistical fluids, fiber suspensions, polymers suspensions, viscoelastic fluids, we see that there are two levels of flow description:

(i) The first one is a level related to the global flow kinematics variables (such as velocity) requiring a variational formulation of the problem overall the physical domain.
(ii) The second one is a level related to the elementary representative volume which characterizes a state of matter: an orientation or conformation induced by the flow. This condition defines the effect of the microscopic structure of the flow.

Taking into account the state of microscopic structure can be done at different scales. One can use in a first approach a constitutive law (differential or integral) to describe the evolution of the stress tensor characterizing the structure. For some behaviors such as fibers, the constitutive law is written in terms of an approximation introducing a closure relation. Unfortunately, in most cases (excluding some constitutive equation written rigorously), there is no equivalence between the definition of the microscopic structure evolution and the constitutive law. In fact, the approximation introduces some errors. This is particularly the case for fiber suspension where there is a high incidence of the error induced by a closure relation on the orientation tensor when the diffusion parameter is small.

If we wish to describe directly the evolution of the structure, a fine modeling at the microscopic scale reveals itself indispensable.

The way in which we describe a microscopic behavior is based on (i) the kinematics of each particle and (ii) the evolution of a probability distribution on the configuration space of all the particles, also called the probability space. From a probability distribution one can go back to the macroscopic state through a calculation of the stress tensor giving the microscopic contribution. We consider that the kinematics of each particle is given by a hydrodynamic contribution and interactions efforts contribution. Terms arising from Brownian effects are obviously taken into account in the diffusive contribution of the convection-diffusion equation characterizing the evolution of the probability distribution. This equation is the so-called Fokker-Planck equation.

For example, in the context of multi-dumbbells models some contributions have allowed to find a solution of the Fokker-Planck equation as a sum of functions products in the context of the PGD [9]. This also has been done in the case of polymer melts [10].

Such approaches of numerical modeling at small scales also have many advantages in the bio-medical field. Macromolecules such as DNA chains can be modeled with high-dimensional configuration spaces. The difficulty arises in situations in which one wants to lead a macromolecule (pharmaceutical drug for example) in into a pipe of very small size (e.g., a vein) without the use of tools. We should then be able to predict the properties of the velocity field so that the macromolecule gets the desired state.

In the same framework, the kinetic description of the rheology of carbon nanotubes suspensions where the direction and also the aggregation state of the system has been addressed [11].

In carbon nanotubes suspensions one must distinguish the case where the nanotubes are functionalized to prevent their aggregation and the case if they are not. This latter situation is able to lead to their aggregation with significant effects on the rheological properties.

In the case of the functionalized nanotubes, kinetic model has been developed for short suspensions and has been relevant to describe the nonlinear rheology.

When now we come back to the upper scale of the global flow (in a framework of micro-macro approach), the difficulty lies in predicting the state of micro-structure which affects directly the final properties. We then have a level of modeling on the scale of the geometry of the flow giving kinematics, thermal field, pressure . . . and a level of modeling of the microscopic characterization, of the state of orientation, of a fiber suspension, or of the conformation of a macromolecule's population . . .

To this end, developments are necessary:

- To be able to consider the relevant microscopic information in order to integrate more physical responses such those finely described with molecular dynamics.
- To integrate the micro-macro coupling—for which we must create the required techniques in adequacy with a rapid integration of microscopic behavior (finely described

at the representative elementary volume of flow with the probability distribution) in a simulation code.

The scope of application of this work is the engineering of complex fluids. Although several studies have been done by substituting the microscopic description by using approximations based on constitutive differential or integral equations, it turns out that the kinematics of the flow is highly affected by the topology of the microstructure; consequently, we have to treat more carefully the microscopic information. The objective is to make the interaction between the kinematics of the flow behavior and the molecular information at the lowest numerical cost. This then requires an appropriate use of specific techniques of model reduction to adequately describe the probability distribution on a hyperspace resulting from a combination of physical space, the configuration space and the temporal dimension.

Conflicts of Interest: The authors declare no conflict of interest.

References

1. Keunings, R. *Micro-Macro Methods for the Multiscale Simulation of Viscoelastic Flow Using Molecular Models of Kinetic Theory, Rheology Reviews*; Binding, D.M., Walters, K., Eds.; British Society of Rheology: London, UK, 2004; pp. 67–98.
2. Advani, S.G.; Tucker, C.L., III. Closure approximations for three-dimensional structure tensors. *J. Rheol.* **1990**, *34*, 367–386. [CrossRef]
3. Ottinger, H.C. *Stochastic Processes in Polymeric Fluids: Tools and Examples for Developing Simulation Algorithms*; Springer: Berlin, Germany, 1996.
4. Öttinger, H.C.; Laso, M. *Smart Polymers in Finite Element Calculation*; International Congress on Rheology: Brussel, Belguim, 1992.
5. Bungartz, H.J.; Griebel, M. Sparse grids. *Acta Numer.* **2004**, *13*, 147–269. [CrossRef]
6. Achdou, Y.; Pironneau, O. *Computational Methods for Option Pricing. SIAM Frontiers in Applied Mathematics*; SIAM: Philadelphia, PA, USA, 2005.
7. Beylkin, G.; Mohlenkamp, M. Algorithms for numerical analysis in high dimensions. *SIAM J. Sci. Comput.* **2005**, *26*, 2133–2159. [CrossRef]
8. Ladeveze, P. *Nonlinear Computational Structural Mechanics*; Springer: New York, NY, USA, 1999.
9. Ammar, A.; Mokdad, B.; Chinesta, F.; Keunings, R. A New Family of Solvers for Some Classes of Multidimensional Partial Differential Equations Encountered in Kinetic Theory Modeling of Complex Fluids: Part II: Transient Simulation Using Space-Time Separated Representation. *J. Non-Newton. Fluid Mech.* **2007**, *144*, 98–121. [CrossRef]
10. Mokdad, B.; Pruliere, E.; Ammar, A.; Chinesta, F. On the Simulation of Kinetic Theory Models of Complex Fluids Using the Fokker-Planck Approach. *Appl. Rheol.* **2007**, *17*, 26494. [CrossRef]
11. Ma, A.; Chinesta, F.; Ammar, A.; Mackley, M. Rheological Modelling of Untreated Carbon Nanotube Suspensions in Simple Steady Shear Flows. *J. Rheol.* **2008**, *52*, 1311–1330. [CrossRef]

Article

Viscosity and Rheological Properties of Graphene Nanopowders Nanofluids

Abderrahim Bakak [1], Mohamed Lotfi [2], Rodolphe Heyd [3,*], Amine Ammar [3] and Abdelaziz Koumina [1]

- [1] Laboratoire Interdisciplinaire de Recherche en Bioressources, Énergie et Matériaux (LIRBEM), ENS, Cadi Ayyad University, Marrakech 40000, Morocco; abderrahim.bakak@gmail.com (A.B.); koumina@uca.ac.ma (A.K.)
- [2] Materials, Energy & Environment Laboratory (LaMEE), FSSM, Cadi Ayyad University, Marrakech 40000, Morocco; lotfimohamed_1999@yahoo.fr
- [3] Laboratoire Angevin de Mécanique, Procédés et innovAtion (LAMPA), Arts et Métiers ParisTech, Boulevard du Ronceray 2, BP 93525, CEDEX 01, F-49035 Angers, France; Amine.AMMAR@ensam.eu
- * Correspondence: Rodolphe.HEYD@ensam.eu

Abstract: The dynamic viscosity and rheological properties of two different non-aqueous graphene nano-plates-based nanofluids are experimentally investigated in this paper, focusing on the effects of solid volume fraction and shear rate. For each nanofluid, four solid volume fractions have been considered ranging from 0.1% to 1%. The rheological characterization of the suspensions was performed at 20 °C, with shear rates ranging from 10^{-1} s^{-1} to 10^3 s^{-1}, using a cone-plate rheometer. The Carreau–Yasuda model has been successfully applied to fit most of the rheological measurements. Although it is very common to observe an increase of the viscosity with the solid volume fraction, we still found here that the addition of nanoparticles produces lubrication effects in some cases. Such a result could be very helpful in the domain of heat extraction applications. The dependence of dynamic viscosity with graphene volume fraction was analyzed using the model of Vallejo et al.

Keywords: graphene nano-powder; thermal nanofluid; rheological behavior; Carreau nanofluid; lubrication effect; Vallejo law

1. Introduction

Global warming and environmental disasters are current events that demonstrate the urgency of a better consideration of renewable energy sources. According to the International Energy Agency (IEA), during 2018, the fossil fuel share represented 81% of the 14,314 Mtoe of the world's primary energy demand, while the share of renewable energy was only 9.7%. According to the IEA, improving energy efficiency is the central factor that will enable the world to move towards a sustainable development scenario. Unfortunately, the IEA also noted a clear slowdown in global progress on energy efficiency in its 2019 report [1], which is of serious concern in the objective to meet global climate targets and other sustainable energy goals. It is therefore vital to improve energy efficiency at all levels of fossil resource use and consequently every reliable contribution in this direction is welcome.

Heat transfer plays a vital role in many industrial and technical applications, ranging from cooling of heat engines or high-power transformers to heat exchangers used in hot water solar panels, refrigeration systems, or power plants. Unfortunately, usual heat transfer fluids (HTFs) such as water (Wa), thermal oils (TOs), ethylene-glycol (EG), or lubricating oils (LOs) all have thermal conductivity less than one unity ($k < 1$ W· m^{-1}·K^{-1}), and this is a significant obstacle in improving the efficiency in thermal energy transfer or extraction.

According to Fourier's law $\mathbf{j}_Q = -k\boldsymbol{\nabla} T$, increasing the thermal conductivity k of HTFs will result in increasing the conductive heat flux between solids and HTFs. Thus, one way to improve heat extraction is to combine the flow properties of HTFs with the high thermal conductivity of some solid materials, such as metals (Cu, Ag, Fe, etc.), metal oxides (CuO, Cu_2O, SiO_2, TiO_2, Al_2O_3, etc.), or different carbon-based materials (carbon black (CB), carbon nanotubes (CNT), and nanohorns (CNH) or graphene (GR)).

However, the use of suspensions with micrometer-sized solid materials (microcomposites) would lead to many prohibitive problems, such as abrasion, sedimentation, and high risk of clogging. Fortunately, advances in nanotechnology now make it possible to synthesize a wide variety of highly thermally conductive solid nanoparticles (NPs), which can be stably suspended in HTFs to form nano-composites (nanofluids and nanolubricants) and impart interesting thermal properties for heat extraction, without the disadvantages mentioned above.

Nanofluids [2] are colloidal suspensions composed of solid nanoparticles (NPs), having at least one dimension that is nanometric in size (<100 nm), stably suspended in a thermal liquid such as water, ethylene glycol, or thermal oils [3–5]. Lubricating oil-based suspensions are also sometimes called nanolubricants [6]. The amazing thermal properties of nanofluids have been the subject of intense investigations in recent years [3,4,7–9]. The potential applications of these nano-suspensions are multiple and promising in various fields, such as cooling power electronic components, industrial and domestic air conditioning and cooling, heat extraction, and transport. As mentioned previously, these suspensions could constitute, under certain mechanical conditions of use which will be discussed later, a promising outlet for nanosciences in the field of energy saving [9–11].

Due to the very large contact areas provided by porous media [12,13], their use in heat exchangers could also be an interesting way to improve heat transfer (in ducts and pipes, for example). It could therefore be possible to combine the two aspects (thermal nanofluids and porous media) to further intensify heat extraction [14], provided of course that the addition of nanoparticles does not significantly increase the base fluid viscosity.

Nanofluids also have a wide range of applications in many other fields than thermal transfers. We can mention, for example, the very promising field of nanomedicine, where nanocarriers are used to allow the delivery of therapeutic and/or imaging agents directly to tumor cells [15,16].

Different kinds of nano-particles (NPs) have been considered so far to produce nanofluids. They can be prepared from polymeric, metallic, organic, and inorganic materials, in the form of tubular (e.g., carbon nanotubes), spherical (metals and oxides) or layered (graphene) structures [10,17–23]. Among these various materials, graphene is a very promising candidate because of its exceptional physical properties, including: a high value of charge carrier mobility [20], exceptional transport performances [22], high specific surface area [24], high thermal conductivity [18], and a significant Young's modulus [25]. These properties rank this allotropic variety of carbon in the category of the most suitable materials for the preparation of thermal nanofluids, which are sought after mainly to improve heat extraction capacities.

While the most cited carbon-based nanomaterials for nanofluids applications are carbon nanotubes [26–31], recently other structures (such as graphene nanoplatelets (GNPs) and reduced graphene oxide (RGO)) have become more widespread [32]. Several researchers have studied the rheological properties of different graphene based nanofluids [9,32–36]. Monireh et al. [35] examined the impact of several parameters on the rheological properties of glycerol and multilayer graphene nanofluids. Their results show that the viscosity increases with the raise in the solid mass fraction (between 0.0025 and 0.0200). They reported an increase in the viscosity (401.49%) of glycerol for 2% graphene nano-sheets fraction, at shear rates of 6.32 s^{-1} and at a temperature of 20 °C. In addition, Kole et al. [34] examined and evaluated the effect of graphene nano-sheets, added to the base fluid (distilled water + ethylene glycol). Their results showed a non-Newtonian behavior with the appearance of a reduction in viscosity by shearing, and an increase of 100% compared to the basic fluid for a graphene volume fraction of 0.395%. In their paper, Kazemi et al. [36] examined the effect of adding Silica and Graphene nanoparticles (using volume fractions 0.05%, 0.1%, 0.25%, 0.5%, and 1%) on viscosity of water. Their experimental results revealed non-Newtonian pseudoplastic behavior of Graphene/water nanofluids. Ahammed et al. [37] have studied the effect of volume fraction (0.05–0.15%) and the temperature (10–90 °C) on the viscosity of nanofluids containing graphene nanosheets dispersed in water. They found that the nanofluid water–graphene viscosity decreases with increasing temperatures and increases with the volume fraction of the nanosheets. An average increase of 47% in viscosity has been noted for 0.15% volume fraction of graphene at 50 °C.

While, from a heat transfer point of view, thermal conductivity is an essential property of thermal nanofluids, from a practical point of view, the dynamic viscosity of these suspensions is an essential property for applications involving fluid flow, as heat transfer and mechanical efficiency are deeply impacted by the viscosity of the fluid [27,32,38]. The addition of nanoparticles to a base fluid can significantly alter its rheological properties, inducing, for example, non-Newtonian behaviors, and, moreover, it can lead to a significant increase of head losses. These pressure losses and rheological behavior alterations can represent a serious limitation to the industrial use of thermal nanofluids [32]. It is therefore important to study them systematically in order to predict the best operating ranges of the considered thermal nanofluids. This is thus the main motivation of the present research.

The improvement of the thermal conductivity and the modification of the viscosity of nanofluids strongly depend on several parameters, among which the size and concentration of the nanoparticles, their nature and shape, the nature of the base fluid, the operating temperature and the shear rate [32,33,39,40].

This paper presents an experimental study of the rheological properties of two thermal nanofluids based on an allotropic variety of graphene nanoparticles, called graphene nano-platelets (GNPs). In this study, the GNPs nanoparticles were dispersed in two kinds of base fluids, with quite different viscosities: an industrial lubricating oil (LO) and ethylene glycol (EG). We are concerned here with the study of the influence of solid particles concentration and of shear rate on the rheological behavior of the suspensions under investigation. Preparation and characterization of the suspensions used in the study are presented in the first part of the paper (Experimental Methods). Then, experimental results are presented and analyzed in terms of the influence of solid volume fraction and shear rate on the rheological properties of the two graphene-based suspensions (Results and Discussion). The experimental results are compared with different models (Carreau–Yasuda and Cross in regard to the shear rate and Vallejo and Maron–Pierce for the solid volume fraction). Conclusions and perspectives for future investigations are finally proposed (Conclusions and Perspectives).

2. Experimental Methods

2.1. Materials

Ethylene glycol (EG) (Sigma-Aldrich, BioUltra $\geq$ 99.5%) and a lubricating oil (LO) (Fuchs, ISO VG 68 RENEP CGLP) were used as base fluids. Table 1 shows the measured η_{mea} and reference η_{ref} viscosity values of pure base liquids at 20 °C. The reference viscosity of EG is given by [41], while the reference value for LO is given by the manufacturer Fuchs. The uncertainty ε_η between measured and reference values is also given in % in Table 1.

Table 1. Measured η_{mea}, reference η_{ref} viscosity values (in mPa·s) and uncertainty ε_η (in %) of pure base liquids at 20 °C.

Base Liquid	η_{ref}	η_{mea}	ε_η
Ethylene Glycol (EG)	19.9	20.3	2.0
Lubricating Oil (LO)	195	187	4.3

The GNPs were purchased from the Graphene Supermarket company (GSc) and used as is. The dry powder of GNPs has a black color, a purity of 99.2%, and a density

$\rho = 2.25$ g·cm^{-3} (from GSc). Figure 1 shows the scanning electron microscope (SEM) images of the GNP sheets (performed at LAMPA with a Zeiss Supra 25 microscope, allowing a 1.5 nm resolution at 20.00 kV). These images clearly show the structural morphology of GNPs in the form of nano-sheets with an average thickness of 3 nm (according to GSc). These sheets are thus composed of 3 to 8 graphene mono-layers. Figure 1 also shows that the nano-sheets are aggregated and overlap randomly.

Figure 1. SEM images of GNP nano-sheets, made at a working distance (WD) of 9.4 mm, an electron high tension (EHT) of 15.00 kV and noise reduction by line average filtering.

The nano-powders were analyzed by Raman spectroscopy, a technique that is commonly used to characterize graphitic materials. Figure 2a shows a typical Raman spectrum obtained with the powders used in this study. The three main peaks characteristic of graphene-based materials are present, with usual relative intensities and widths: G (∼1580 cm^{-1}) and 2D (∼2690 cm^{-1}) peaks that are always present in the case of graphene and the D peak (∼1350 cm^{-1}), which indicates the presence of defects [42]. Raman spectra were performed using a Confotec MR520 Raman spectrometer at $\lambda = 532$ nm, with an analysis time of 30 s.

X-ray diffraction (XRD) analysis was also conducted on the nano-powders. Figure 2b shows that the diffraction pattern of graphene powders presents two main peaks at $2\theta \sim 27°$ and $2\theta \sim 54°$, which are very close to the diffraction peaks of graphite, as mentioned in [43].

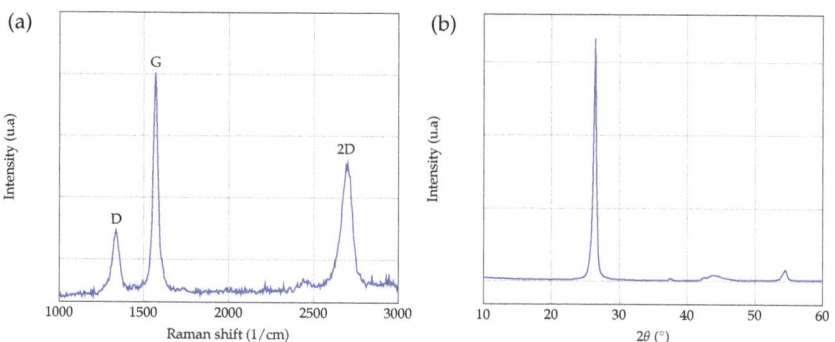

Figure 2. Raman spectrum (**a**) and XRD spectrum (**b**) of the GNPs powder used in this study.

2.2. Graphene Suspensions

Different masses of GNPs were dispersed in 20 mL of each base fluid, to obtain the following solid volume fractions: $\phi = 0.1\%$, 0.25%, 0.5%, and 1%. No dispersing agent or surfactant has been used in the formulation. Each mixture has been stirred with a magnetic stirrer for 48 h, to ensure a uniform dispersion of the nano-particles in the base fluid. In order to limit the initial agglomeration of the nano-particles in the base fluid, the solutions were exposed to moderate sonication (LEO ultrasonic bath, oscillation frequency 46 kHz and power 80 W) for 2 h. The samples, contained in closed vials, were immersed in a water bath at room temperature. No significant variation of the samples temperature was observed during the sonication process. Next, all suspensions were stored at room temperature in hermetic containers. No observable phase separation has been detected before and after rheological measurements. In all the cases considered here, the preparation of suspensions with a solid volume fraction of 2% has led to samples that were no longer liquid but rather pasty and that behaved like gels (no flow under the effect of gravity, by turning the vial upside down). This study is therefore limited to solid volume fractions of less than 2%.

2.3. Suspensions Characterization

To study the state of dispersion of the nano-particles in the base fluid and to evaluate the presence and size of graphene aggregates [33,44–47], samples of each suspension were analyzed by SEM after drying (Figure 3 shows an example of nanofluid based on ethylene glycol, for a solid volume fraction $\phi = 0.25\%$: EG-GNPs-0.25). One drop of each sample was collected, placed on the SEM grid and then slowly dried in an oven (see Table 2). Figure 3 shows that the graphene nano-sheets are uniformly dispersed, revealing no irreversible agglomerates, and that the morphology of these nano-sheets is not noticeably altered after the stirring and sonication steps.

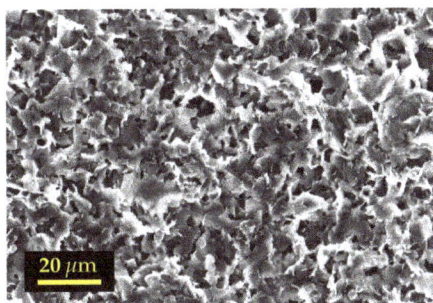 

Figure 3. SEM characterization of EG-GNPs-0.25 nanofluid.

Table 2. Characteristics of oven sample processing.

Base Fluid	EG	LO
Oven duration	12 h	72 h
Temperature	100 °C	220 °C

The rheological study of the nanofluids was carried out with a rotational rheometer (Malvern Kinexus Pro), using a cone-plate geometry (1°–60 mm), temperature-controlled with a resolution of 0.01 °C. The geometry and the liquid to be characterized were enclosed under a cover (hood), in order to improve the temperature homogeneity within the sample. All the dynamic viscosity measurements were performed with the same geometry. No particular experimental problems, such as material rejection or phase separation, were observed during the measurements.

2.4. Rheological Properties of Base Liquids

The first experiments were carried out on the base liquids, in order to evaluate the uncertainty of the rheometer. All measurements were repeated at least twice to check their reproducibility. Figure 4 shows the results of our measurements obtained for pure ethylene glycol, the less viscous base liquid used in the present study, at three working temperatures

(20.00 °C, 40.00 °C, and 60.00 °C). The values that we have measured for these three temperatures are quite close to those obtained by Chen et al. [48] and Sawicka et al. [41]. As expected, the dynamic viscosity (η) of ethylene glycol is independent of shear rate. Similar results were obtained for the lubricant oil. The base fluids used in this study behave like Newtonian fluids over the whole temperature range under investigation. The relative measurement uncertainty has been estimated to be on the order of 2% at 20.00 °C.

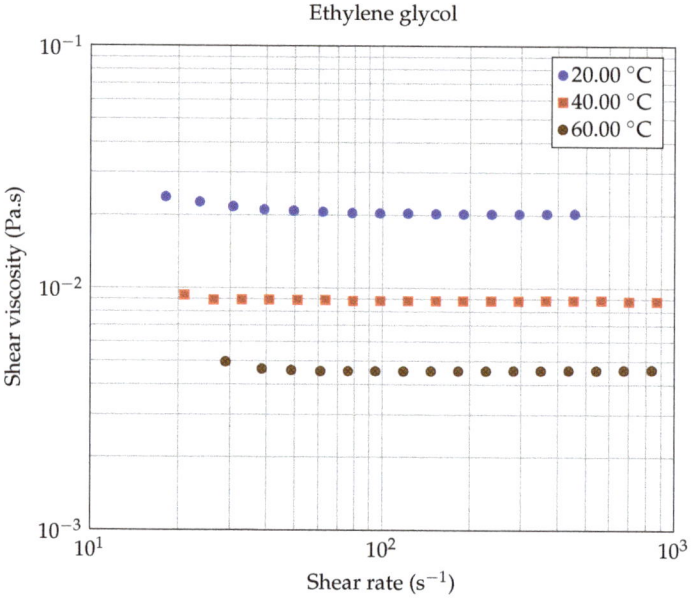

Figure 4. Rheological behavior of ethylene glycol at different working temperatures.

Figure 5 shows the evolution of dynamic viscosity as a function of temperature for each of the two base fluids used in our study. As expected, it is observed that the dynamic viscosity η_{bf} of these base fluids is a decreasing function of temperature, according [48] to usual laws of type:

$$\ln \eta_{bf} = A + 1000 \times \frac{B}{T+C}, \quad (1)$$

where T is the absolute temperature, η_{bf} is the base fluid viscosity (in mPa·s), and A, B, and C are fluid specific constants. Even more specific laws can be used (see, for example, Bird et al. [49], chapter 1). It is also possible to write (1) in the so-called Vogel–Fulcher–Tamman form [50,51]:

$$\eta_{bf} = \eta_0 e^{\frac{DT_0}{T-T_0}}, \quad (2)$$

where η_0 (in Pa.s), T_0 (in K) and D are fitting parameters related to A, B, and C by the following relations: $\eta_0 = 10^{-3} \cdot e^A$, $T_0 = -C$ and $D = -10^3 \cdot B/C$.

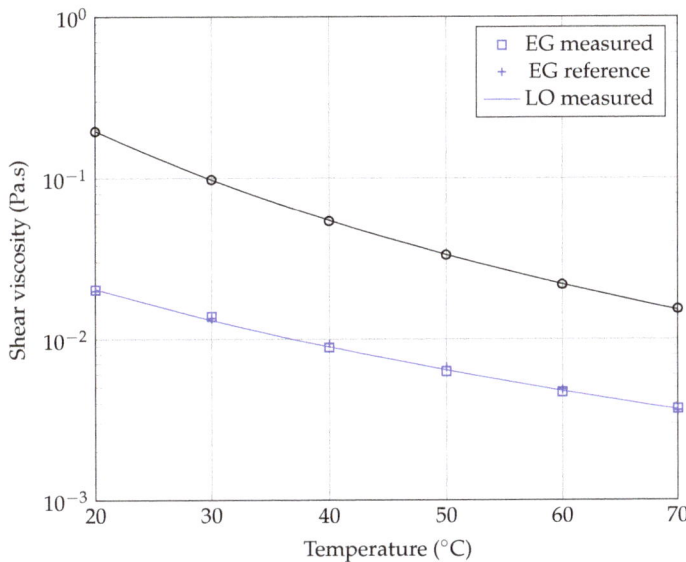

Figure 5. Evolution of the dynamic viscosity of base liquids as a function of temperature. The continuous lines represent model (1) with the respective coefficients of Table 3.

Table 3 gathers the values of coefficients A, B, C and of the determination coefficient R^2 calculated for the two base fluids used in this study.

Table 3. A, B, and C coefficients of Equation (1) and η_0, T_0, and D coefficients of Equation (2), calculated for the two base fluids used in this study.

Liquids	A	B	C	$\eta_0/10^{-5}$	T_0	D	R^2
EG	−3.202	0.813	−162.5	40.68	162.5	5.003	0.9974
LO	−2.353	0.757	−194.0	95.08	194.0	3.902	0.9998

In [41], Sawicka et al. used an Arrhenius-type law to model their measurements of EG viscosity as a function of absolute temperature T:

$$\eta_{EG} = A \exp\left(\frac{B}{T}\right), \tag{3}$$

where $A = 1.6 \times 10^{-7}$ Pa.s and $B = 3440$ K. Note that, using model (3), our experimental measurements led to the following values in the case of EG: $A = 1.11 \times 10^{-7}$ Pa·s and $B = 3548$ K, with a coefficient of determination $R^2 = 0.9971$. These results confirm the consistency of the present viscosity measurements with those of Sawicka et al. Table 4 collects our measurements η_{mea} of viscosity versus temperature for ethylene glycol and compares them with the results η_{ref} obtained by Sawicka et al. in [41]. The corresponding uncertainties ε_η are also given in %.

Table 4. Ethylene Glycol viscosity values (in Pa·s) as a function of temperature.

T (°C)	20.00	30.00	40.00	50.00	60.00	70.00	80.00
η_{mea} (Pa·s)	0.0203	0.0139	0.0089	0.0063	0.0047	0.0037	0.0029
η_{ref} (Pa·s)	0.0199	0.0135	0.0094	0.0067	0.0049	0.0036	0.0027
$|\varepsilon_\eta|$ (%)	2.0	2.8	5.3	6.0	4.1	2.0	5.8

3. Results and Discussion

3.1. Rheological Behavior

Figures 6 and 7 show the rheological behavior of the two nanofluids for different volume fractions, at a working temperature of 20.00 °C, and as a function of the shear rate. Within the shear rates range investigated, it is observed that the rheological behavior of the nanofluid is strongly dependent on the solid volume fraction. For each of the two studied nanofluids, shear-thinning has been observed, which is more pronounced for higher solid volume fractions. As indicated in the literature [35], the decrease in viscosity as a function of shear rate could be attributed to a de-agglomeration effect of the graphene nanosheets or to the alignment of the nanosheets in the plane of flow during shearing [48], resulting in less viscous dissipation and consequently in a decreasing of the apparent viscosity of the suspension. As the volume fraction of GNPs suspended in the base fluid increases, shear-thinning deviation from Newtonian behavior becomes more and more pronounced. Very similar behaviors were observed by Vallejo et al. [51] using nanofluids composed of an ethylene-glycol:water mixture (50:50 vol%) and different carbon-based nanomaterials (carbon black, nanodiamonds, graphite/diamond mixtures and sulfonicacid-functionalized graphene nanoplatelets).

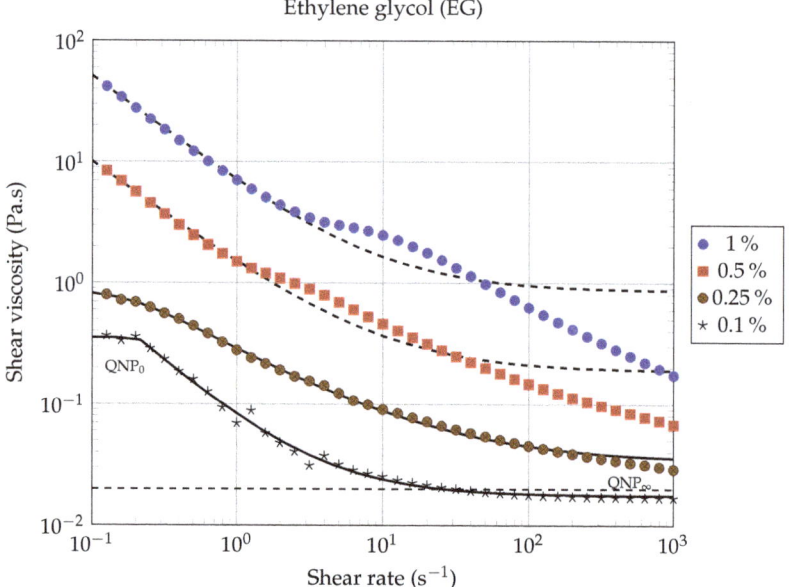

Figure 6. Dynamic viscosity of EG-GNP nanofluids as a function of shear rate, for different solid volume fractions, at 20.00 °C. Solid and dashed lines correspond to the model (6), using the coefficients gathered in Table 5. The horizontal dashed line indicates the viscosity value of the base fluid.

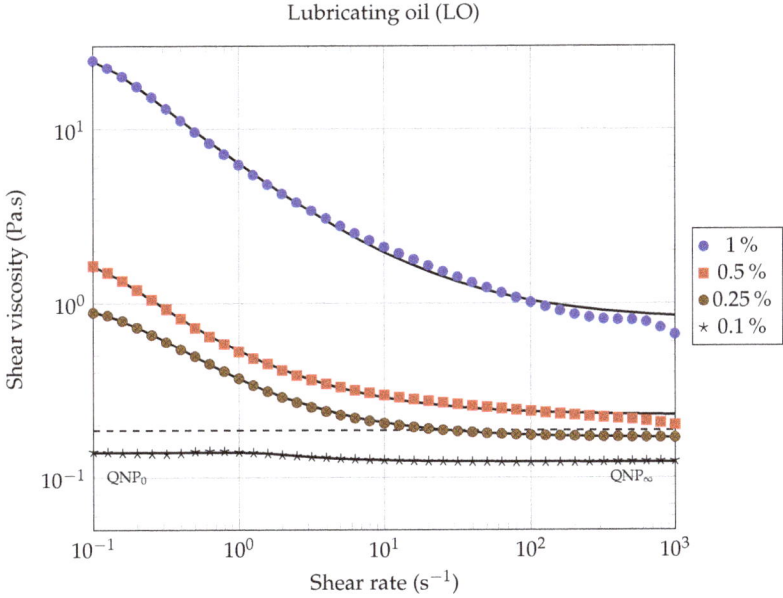

Figure 7. Dynamic viscosity of LO-GNP nanofluids as a function of shear rate, for different solid volume fractions, at 20.00 °C. Solid lines correspond to the model (6), using the coefficients gathered in Table 5. The dashed line indicates the viscosity value of the base fluid.

Table 5. Values of the Carreau–Yasuda parameters η_0, η_∞, a, λ, and n obtained by fitting the experimental results. BF means base fluid (EG: Ethylene Glycol and LO: Lubricating Oil) at (@) solid volume fraction ϕ (in %). R^2_{CY} and R^2_{CM} are the determination coefficients corresponding to the Carreau–Yasuda and the Cross models, respectively. R^2_{PL} is the determination coefficient of the power law model, which has been applied only for $\phi = 0.1\%$.

BF@ϕ	η_0 (Pa·s)	η_∞ (Pa·s)	a	λ	n	R^2_{CY}	R^2_{CM}	R^2_{PL}
EG@0.1	0.3535	1.7574×10^{-2}	123.84	4.8878	−0.0216	0.9967	0.9910	0.9689
EG@0.25	0.8731	3.3330×10^{-2}	2.3583	5.7222	0.3245	0.9992	0.9984	–
EG@0.5	19.776	0.1871	21.873	21.731	0.1295	0.9969	0.9969	–
EG@1.0	117.11	0.8536	18.701	25.214	0.0981	0.9978	0.9978	–
LO@0.1	0.1395	0.1229	7.3295	0.6650	0.0861	0.9753	0.9696	0.8217
LO@0.25	1.0120	0.1683	1.9802	7.1384	0.2703	0.9999	0.9996	–
LO@0.5	1.7795	0.2264	4.8753	9.7821	0.2927	0.9995	0.9989	–
LO@1.0	27.642	0.8118	4.1630	9.9430	0.3159	0.9998	0.9992	–

For each of the two types of nanofluids studied here, two quasi-Newtonian plateaus can be observed for the lowest solid volume fraction ($\phi = 0.1\%$). The first one, denoted QNP$_0$, is observed at low shear rates (see Figures 6 and 7), while the second one, denoted QNP$_\infty$, is observed at high shear rates. The QNP$_\infty$ plateau is also present for the volume fraction ($\phi = 0.25\%$), but only in the case of LO based nanofluids. It can be noted that the extent of each of these plateaus depends on both the nature of the base liquid considered and on solid volume fractions. The presence of such Newtonian plateaus in the rheological behavior of nanofluids based on carbonaceous nanomaterials has also been reported by Vallejo et al. [51].

In the absence of Newtonian plateaus, the shear-thinning behavior of suspensions is often well described using a power law ([52] chapter 5, page 90), also known as the Ostwald–de Waele law (PL):

$$\eta = k|\dot{\gamma}|^{n-1}, \qquad (4)$$

where $\dot{\gamma}$ is the shear rate, $k > 0$ is the flow consistency index, and n is the flow behavior index ($n < 1$ for shear-thinning behavior). From the results shown in Figures 6 and 7, it is clear that a power law of type (4) cannot describe the whole contour shape of the flow curves for the different nanofluids studied here. It can only describe a small range of shear rates, corresponding to the shear thinning region. We have illustrated the inability of the power law (PL) to adequately describe the whole rheological behavior of our nanofluids, firstly in Figures 8 and 9, where we have compared different rheological models in the case of the lowest solid volume fraction $\phi = 0.1\%$ and secondly, in Table 5, where the corresponding coefficient of determination R^2_{PL} has been calculated when applying the power law (4) over the whole shear rates domain.

In [51], Vallejo et al. used the Cross model (CM) to fit their rheological measurements:

$$\eta = \eta_\infty + \frac{\eta_0 - \eta_\infty}{1 + (k \cdot \dot{\gamma})^m}, \qquad (5)$$

where m and k are called the rate constant and the time constant, respectively, while η_0 and η_∞ are the asymptotic values of dynamic viscosity corresponding to QNP_0 and QNP_∞, respectively [51]. This law has shown to be more suitable than the power law to describe our measurements over the entire range of shear rates studied (see Figures 8 and 9).

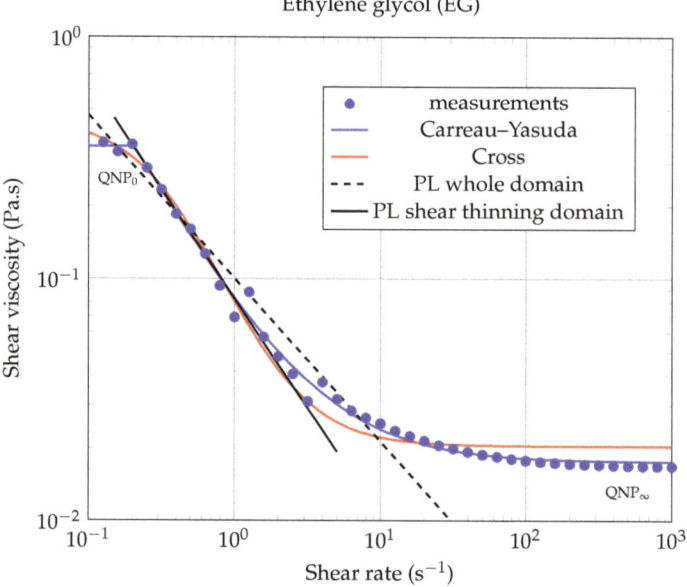

Figure 8. Dynamic viscosity of EG-GNP nanofluid as a function of shear rate, for the lowest solid volume fraction $\phi = 0.1\%$, at 20.00 °C. The Carreau–Yasuda model (6) has been plotted using the coefficients gathered in Table 5. In the shear thinning domain, PL modeling led to a quite good coefficient of determination $R^2_{PL} = 0.9981$.

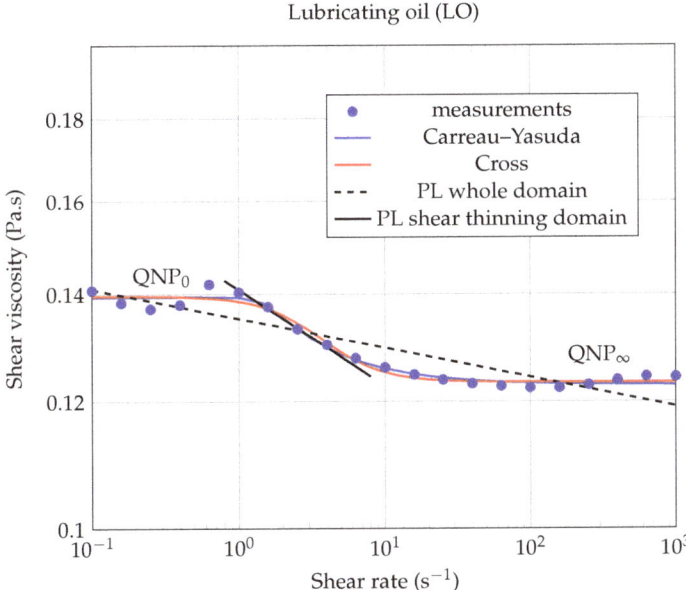

Figure 9. Dynamic viscosity of LO-GNP nanofluid as a function of shear rate, for the lowest solid volume fraction $\phi = 0.1\%$, at 20.00 °C. The Carreau–Yasuda model (6) has been plotted using the coefficients gathered in Table 5. In the shear thinning domain, PL modeling led to a coefficient of determination $R_{PL}^2 = 0.9891$.

Our experimental data were also fitted using the Carreau–Yasuda (CY) model for shear-thinning fluids [53,54]:

$$\frac{\eta - \eta_\infty}{\eta_0 - \eta_\infty} = \left[1 + (\lambda \dot{\gamma})^a\right]^{\frac{n-1}{a}}, \quad (6)$$

where η_0 is the zero shear rate dynamic viscosity (corresponding to QNP$_0$); η_∞ is the high share rate dynamic viscosity (corresponding to QNP$_\infty$) and, according to Kowalska et al., λ is a relaxation time characteristic of the studied fluid and a is a parameter characteristic of the transition width between the zero shear rate viscosity domain and the shear thinning domain. The values of the Carreau–Yasuda parameters, obtained by fitting our experimental results, are gathered in Table 5 and were used to plot the continuous and dashed lines (except horizontal lines) in Figures 6 and 7. It can be seen from these figures that, at the lowest solid volume fractions used here ($\phi \leq 0.25\%$), the rheological behavior of each of the two types of nanofluids is well described by the CY model, over the whole range of shear rates investigated here.

It should be noted that the CY model consistently gave better results than the Cross model, for each of the nanofluids studied in this work (see the coefficients of determination R_{CY}^2 and R_{CM}^2 collected in Table 5). Therefore, we will discuss hereafter only the results given by the CY model.

It can be noticed that the CY model still applies here remarkably well for the highest solid concentrations studied in this work ($\phi = 0.5\%$ and $\phi = 1.0\%$), but only at low shear rates.

This remark is particularly true in the case of ethylene glycol-based nanofluids (EG-GNPs), where it is observed for shear rates above 2 s^{-1} that the rheological behavior completely fails the Carreau–Yasuda model (see Figure 6, dashed lines). These large deviations from the CY model certainly reflect the increasing influence of graphene–graphene and graphene–ethylene glycol interactions on the rheological properties of the suspension, as the solid volume fraction and the shear rate increase. This very particular rheological behavior has also been observed in some cases by Vallejo et al., for high solid mass fractions (see [51], Figure 3: 0.50wt%Nd97 and Figure 4e: 2.0wt%nD87 and nD97).

On the other hand, we have found that the CY model is particularly well suited for lubricating oil-based nanofluids (LO-GNPs) over the entire shear rates range, as can be seen from the curves plotted in Figure 7 and from the results gathered in Table 5. The value of the coefficient of determination R^2_{CY} is in this case very close to one, for three of the four LO-based nanofluids studied here.

The significant differences in rheological behaviors observed in this work clearly highlight the influence played by base liquid/GNPs interactions on the rheological properties of graphene nanopowder-based suspensions.

3.2. Effect of Solid Volume Fraction on Dynamic Viscosity

We now analyze the influence of suspending various graphene nanosheets volume fractions (ϕ = 0.1%, 0.25%, 0.5% and 1.0%) on the dynamic viscosity of each of the two base fluids under investigation, at the working temperature $T = 293.15$ K.

The dependence of the room temperature dynamic viscosity with solid volume fraction ϕ of EG-GNPs and LO-GNPs nanofluids has been compared to several widely used models, namely Einstein [55], Brinkman [56], Batchelor [57], or Krieger–Dougherty [58] laws (see Table 6). These laws, valid only for spherical nanoparticles, proved to be totally inadequate with the nanofluids studied here, which contain graphene nanosheets.

Table 6. Some models commonly used to estimate the viscosity of micro-dispersions as a function of the solid particles volume fraction. The intrinsic viscosity $[\eta]$ has a typical value of 2.5 for spherical particles, ϕ_m is the maximum particle packing fraction (which has been chosen here as an adjustment parameter) and usually $5.2 \leq \alpha \leq 7.6$.

Models	Einstein	Brinkman	Batchelor	Krieger–Dougherty
$\eta/\eta_{bf} =$	$1 + 2.5\phi$	$1/(1-\phi)^{2.5}$	$1 + 2.5\phi + \alpha\phi^2$	$\left(1 - \frac{\phi}{\phi_m}\right)^{-[\eta]\phi_m}$

The dynamic viscosity η was calculated at 20.00 °C as a function of the solid volume fraction ϕ for each nanofluid, using the experimental results shown in Figures 6 and 7, for the following shear rates: $\dot{\gamma}$ = 0.1, 1.0, 10, 100 and 1000 s^{-1}. The evolution of dynamic viscosity as a function of the GNP volume fraction is shown in Figure 10 for the case of ethylene glycol-based nanofluids and in Figures 11 and 12 in the case of lubricating oil. The dynamic viscosity values of most of the present nanofluids were well modeled using the law proposed by Vallejo et al. [32,59]:

$$\eta = \eta_0 e^{\frac{DT_0}{T-T_0}} + Ee^{\frac{F}{T}} \cdot \phi - G \cdot \phi^2 \qquad (7)$$

where the parameters η_0, D, and T_0, whose values are gathered in Table 3, are specific to the base fluid; E, F, and G are fitting parameters. Since only one temperature T is considered in the present study, we rewrite Vallejo's law in the following simplified form:

$$\eta = \eta_0 e^{\frac{DT_0}{T-T_0}} + E' \cdot \phi - G \cdot \phi^2 \qquad (8)$$

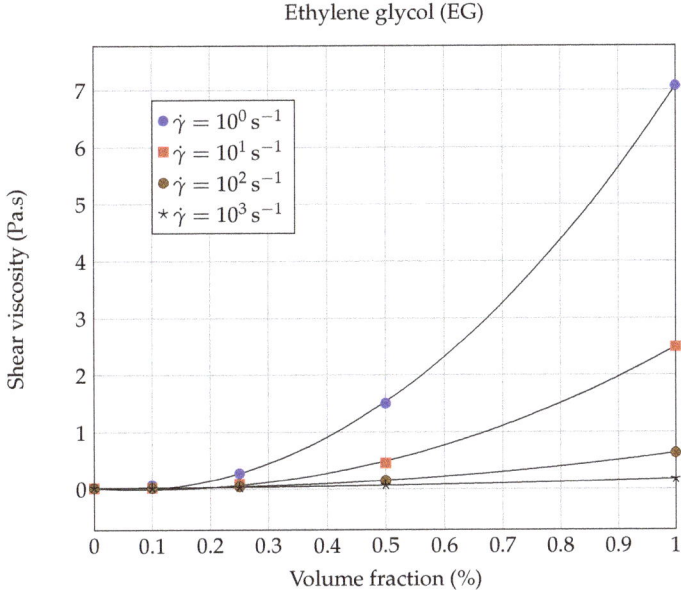

Figure 10. Evolution of the shear viscosity η of EG based nanofluids as a function of the solid volume fraction ϕ, at 20.00 °C and different shear rates. The continuous lines correspond to the Vallejo model (8), with the coefficients collected in Table 7.

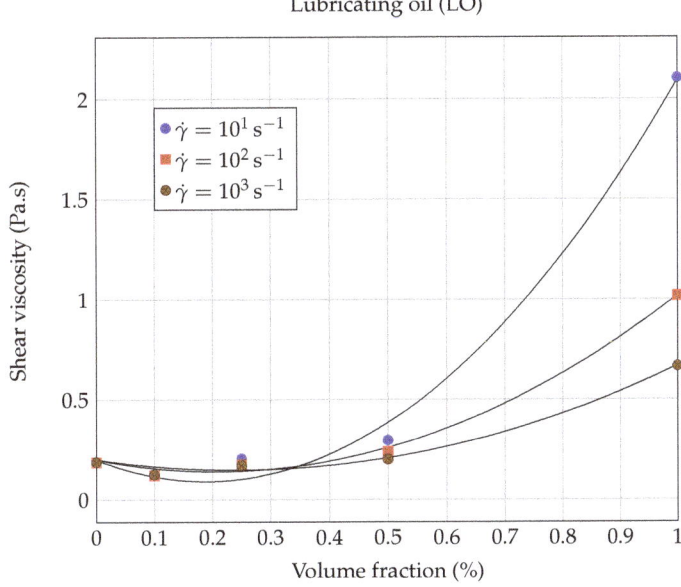

Figure 11. Evolution of the shear viscosity η of LO based nanofluids as a function of the solid volume fraction ϕ, at 20 °C for moderate to high shear rates. The continuous lines correspond to the Vallejo model (8), with the coefficients collected in Table 8.

Table 7. Values of the Vallejo parameters **E'** and **G** obtained by fitting the experimental results of EG based nanofluids, for different shear rates, at 20.00 °C.

Shear Rate (s^{-1})	E' (Pa·s)	G (Pa·s)	R^2
0.1	−8.3824	−50.4425	0.9996
10^0	−0.9938	−8.0442	0.9998
10^1	−0.5826	−3.0493	0.9994
10^2	−0.0908	−0.7015	0.9999
10^3	0.0143	−0.1411	0.9953

Table 8. Values of the Vallejo parameters **E'** and **G** obtained by fitting the experimental results of LO based nanofluids, for different shear rates, at 20.00 °C.

Shear Rate (s^{-1})	E' (Pa·s)	G (Pa·s)	R^2
0.1	−12.723	−37.105	0.9886
10^0	−3.2195	−9.2146	0.9884
10^1	−1.1533	−3.0482	0.9935
10^2	−0.5789	−1.3981	0.9954
10^3	−0.4336	−0.9024	0.9900

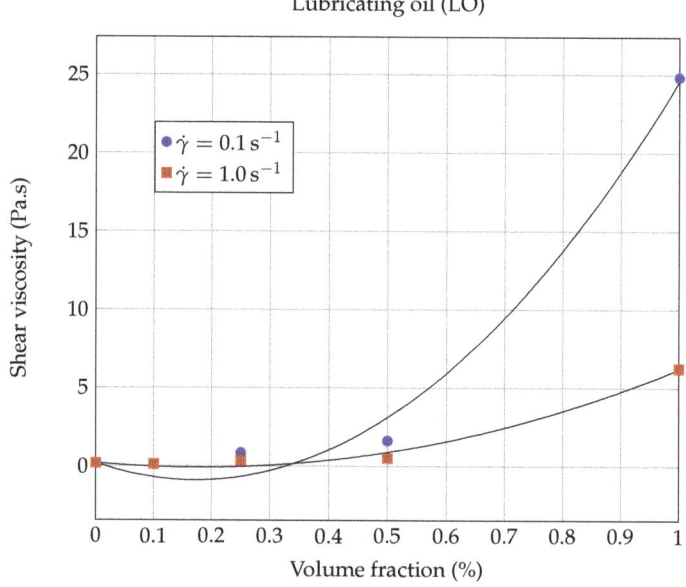

Figure 12. Evolution of the shear viscosity η of LO based nanofluids as a function of the solid volume fraction ϕ, at 20.00 °C and low shear rates $\dot{\gamma} = 0.1\,s^{-1}$ and $\dot{\gamma} = 1.0\,s^{-1}$. The continuous lines correspond to the Vallejo model (8), with the coefficients collected in Table 8.

The Maron and Pierce equation [32] was also used to model our measurements:

$$\eta = \eta_{bf}\left(1 - \frac{\phi}{\phi_m}\right)^{-2} \quad (9)$$

where ϕ_m can be considered as a fitting parameter. This model, which gave good results for aqueous nanofluids containing graphene oxides [60], however, did not provide satisfactory results with the present nanofluids. Therefore, it will not be developed in the rest of this work.

More specifically, in the case of ethylene glycol a quasi systematic increase in the dynamic viscosity with the GNP volume fraction has been observed, whatever the shear rate considered (see Figures 6 and 10); these results are similar to those already published in several studies [34,35,40,51,61–64]. However, we noticed a weak lubricating effect, at the limit of the measurement uncertainty, for the lowest volume fraction considered ($\phi = 0.1\%$) and for shear rates above $40\,\text{s}^{-1}$.

Thus, it can be seen that from the point of view of mechanical performance that it is not very favorable to load ethylene glycol with graphene nanopowders, since it can significantly increase the effective dynamic viscosity of the suspension. This increase is more important as the shear rates considered are low. For example, for $\dot{\gamma} = 1.0\,\text{s}^{-1}$ and $\phi = 1\%$, we found a relative increase in viscosity that is equal to $\eta_r(1\,\text{s}^{-1}) = \eta/\eta_{bf} \approx 350$, which is considerable. Relative increases of this order of magnitude, or even greater, have already been observed in the past with aqueous nanofluids based on carbon nanotubes (CNTs). For example, Ding et al. [65] observed huge variations in dynamic viscosity as a function of shear rate with water-CNT nanofluids. In the case of 0.5 wt.% CNTs suspended in water, they found at $25\,°\text{C}$ that $\eta_r(10^3\,\text{s}^{-1}) \approx 10$ while $\eta_r(1\,\text{s}^{-1}) \approx 10^5$.

The variations of the dynamic viscosity η of the suspensions as a function of the solid volume fraction ϕ were modeled using Vallejo's law (8), at different shear rates $\dot{\gamma}$. As can be seen from the solid line curves shown in Figure 10 and from the values of the coefficients of determination collected in Table 7, it can be noticed that the model of Vallejo et al. applies quite well to the description of the evolution of EG based nanofluids viscosity as a function of graphene volume fraction ϕ, whatever the shear rate considered. These remarkable results confirm the interest of Vallejo's model for ethylene glycol-based nanofluids containing graphene nanopowders.

Next, the viscosity values of the nanofluids based on lubricating oil and graphene nanopowders (LO-GNPs) are analyzed. Figure 7 shows a very interesting and promising rheological behavior, since the addition of GNPs leads here, for the lowest volume fraction $\phi = 0.1\%$, to a decrease of the viscosity compared to the base fluid ($\eta_r(0.1\,\text{s}^{-1}) = \eta/\eta_{bf} = 0.75$ and $\eta_r(1000\,\text{s}^{-1}) = 0.66$), rather than an increase, and this whatever the shear rate considered. Chen et al. [29] have also observed, in the case of a nanofluid prepared with a volume fraction of 0.4% carbon nanotubes (CNT) suspended in EG, that the effective viscosity of the suspension is lower than that of the base fluid, due to a lubrication effect of nanoparticles, which can also be assumed here to be the reason for the viscosity decrease. For higher volume fractions given by 0.5% and 1%, the viscosity of the nanofluid becomes greater than that of the base fluid, whatever the shear rate considered. These observations reflect a non-monotonic behavior of the dynamic viscosity as a function of the solid volume fraction that can be seen in Figure 11. The possibility of a lubrication effect, in the case of the LO-GNP nanofluids for solid volume fractions, which should allow a significant improvement in heat extraction, is a very encouraging result for the use of these liquids from an industrial point of view.

As can be seen from Figure 11 and Table 8, Vallejo's model gave adequate results for moderate ($\dot{\gamma} = 10^1\,\text{s}^{-1}$; $R^2 = 0.9935$) to high shear rates ($\dot{\gamma} = 10^2\,\text{s}^{-1}$; $R^2 = 0.9954$ and $\dot{\gamma} = 10^3\,\text{s}^{-1}$; $R^2 = 0.9900$). On the other hand, the agreement is much less beneficial (see Figure 12) for the two lower shear rates considered here: $\dot{\gamma} = 0.1\,\text{s}^{-1}$ ($R^2 = 0.9886$) and $\dot{\gamma} = 1.0\,\text{s}^{-1}$ ($R^2 = 0.9884$).

In contrast to the case of EG-GNP nanofluids, it can be deduced from the present measurements that the viscosity of LO-GNP nanofluids seems not to verify Vallejo's law correctly at low shear rates. Since the nanoparticles are of the same nature for both types of nanofluids considered in this study, this difference in behavior should probably be attributed to the fluid-GNP interactions.

4. Conclusions and Perspectives

An experimental study of the rheological properties of two different graphene based nanofluids was presented for the following base fluids: ethylene glycol and an industrial lubricating oil. The influence of nanoparticles concentration on the rheological properties of the suspensions has been systematically studied, for different solid volume fractions (0.1%, 0.25%, 0.5% and 1%) at a working temperature of 20.00 °C. The rheological properties of the suspensions were analyzed using both the Carreau–Yasuda and Cross models for shear thinning liquids. For each of the two types of nanofluids considered, the Carreau–Yasuda model gave the best results, the agreement being particularly good at low graphene concentrations ($\phi \leq 0.25\,\%$). However, the presence of graphene at higher concentrations can lead to deviations from the Carreau–Yasuda law, which become more significant at high shear rates.

The Vallejo model was successfully applied to ethylene glycol-based nanofluids whatever the shear rate considered. In the case of lubricating oil-based nanofluids, the dependence of the viscosity on the solid volume fraction is moderately well described by Vallejo's law for low shear rates. Further research will be needed to determine whether other base liquids also escape Vallejo's law for low shear rates in the case of graphene-based nanofluids.

The suspensions studied in this work have exhibited a wide variety of original rheological behaviors. A lubrication effect has been demonstrated for the nanofluid based on lubricating oil, for which the viscosity decreases with the addition of graphene nano-sheets at $\phi = 0.1\,\%$. This interesting behavior allows us to consider industrial applications for this nanofluid in the field of heat extraction, for example, without sacrificing the mechanical performance.

Future work will focus on the rheological behavior of these two types of nanofluids as a function of temperature, but also on the thermal and thermodynamic properties (thermal conductivity, specific heat, solidification temperature) and on the electrical and dielectric properties (electrical conductivity and dielectric permeability).

Author Contributions: Conceptualization, A.B., M.L., R.H., A.A. and A.K.; Investigation, A.B., M.L., R.H., A.A. and A.K.; Methodology, A.B., M.L., R.H., A.A. and A.K.; Project administration, R.H., A.A. and A.K.; Supervision, R.H.; Writing—original draft, A.B., M.L., R.H., A.A. and A.K. All authors have read and agreed to the published version of the manuscript.

Funding: This research received no external funding.

Institutional Review Board Statement: Not applicable.

Informed Consent Statement: Not applicable.

Acknowledgments: The authors would like to thank the members of LAMPA (Laboratoire Angevin de Mécanique, Procédés et innovAtion), for their technical and scientific support during the rheological characterization campaign of the suspensions. The authors would also like to thank the reviewers for their valuable and constructive comments.

Conflicts of Interest: The authors declare no conflict of interest.

References

1. *World Energy Outlook 2019*; International Energy Agency: Paris, France, 2019.
2. Choi, U. *Enhancing Thermal Conductivity of Fluids with Nanoparticles*; Technical Report FED; ASME: New-York, NY, USA, 1995; Volume 231.
3. Akilu, S.; Sharma, K.; Baheta, A.T.; Mamat, R. A review of thermophysical properties of water based composite nanofluids. *Renew. Sustain. Energy Rev.* **2016**, *66*, 654–678. [CrossRef]
4. Murshed, S.S.; Nieto de Castro, C. Conduction and convection heat transfer characteristics of ethylene glycol based nanofluids—A review. *Appl. Energy* **2016**, *184*, 681–695. [CrossRef]
5. Asadi, A.; Aberoumand, S.; Moradikazerouni, A.; Pourfattah, F.; Żyła, G.; Estellé, P.; Mahian, O.; Wongwises, S.; Nguyen, H.M.; Arabkoohsar, A. Recent advances in preparation methods and thermophysical properties of oil-based nanofluids: A state-of-the-art review. *Powder Technol.* **2019**, *352*, 209–226. [CrossRef]
6. Azman, N.F.; Samion, S. Dispersion Stability and Lubrication Mechanism of Nanolubricants: A Review. *Int. J. Precis. Eng.-Manuf.-Green Technol.* **2019**, *6*, 393–414. [CrossRef]
7. Gao, Y.; Wang, H.; Sasmito, A.P.; Mujumdar, A.S. Measurement and modeling of thermal conductivity of graphene nanoplatelet water and ethylene glycol base nanofluids. *Int. J. Heat Mass Transf.* **2018**, 97–109. [CrossRef]

8. Ahmadi, M.H.; Mirlohi, A.; Nazari, M.A.; Ghasempour, R. A Review of Thermal Conductivity of Various Nanofluids. *J. Mol. Liq.* **2018**, *265*, 181–188. [CrossRef]
9. Arshad, A.; Jabbal, M.; Yan, Y.; Reay, D. A Review on Graphene Based Nanofluids: Preparation, Characterization and Applications. *J. Mol. Liq.* **2019**, *279*, 444–484. [CrossRef]
10. Devendiran, D.K.; Amirtham, V.A. A Review on Preparation, Characterization, Properties and Applications of Nanofluids. *Renew. Sustain. Energy Rev.* **2016**, *60*, 21–40. [CrossRef]
11. Yu, W.; Xie, H. A Review on Nanofluids: Preparation, Stability Mechanisms, and Applications. *J. Nanomater.* **2012**, *20*, 1–17. [CrossRef]
12. Liang, M.; Fu, C.; Xiao, B.; Luo, L.; Wang, Z. A fractal study for the effective electrolyte diffusion through charged porous media. *Int. J. Heat Mass Transf.* **2019**, *137*, 365–371. [CrossRef]
13. Zhu, Q.; Xie, M.; Yang, J.; Chen, Y.; Liao, K. Analytical determination of permeability of porous fibrous media with consideration of electrokinetic phenomena. *Int. J. Heat Mass Transf.* **2012**, *55*, 1716–1723. [CrossRef]
14. Kasaeian, A.; Daneshazarian, R.; Mahian, O.; Kolsi, L.; Chamkha, A.J.; Wongwises, S.; Pop, I. Nanofluid flow and heat transfer in porous media: A review of the latest developments. *Int. J. Heat Mass Transf.* **2017**, *107*, 778–791. [CrossRef]
15. David, A. Peptide ligand-modified nanomedicines for targeting cells at the tumor microenvironment. *Adv. Drug Deliv. Rev.* **2017**, *119*, 120–142. [CrossRef]
16. Nguyen, P.V.; Allard-Vannier, E.; Chourpa, I.; Hervé-Aubert, K. Nanomedicines functionalized with anti-EGFR ligands for active targeting in cancer therapy: Biological strategy, design and quality control. *Int. J. Pharm.* **2021**, *605*. [CrossRef]
17. Wang, X.Q.; Mujumdar, A.S. Heat Transfer Characteristics of Nanofluids: A Review. *Int. J. Therm. Sci.* **2007**, *46*, 1–19. [CrossRef]
18. Cao, H.Y.; Guo, Z.X.; Xiang, H.; Gong, X.G. Layer and Size Dependence of Thermal Conductivity in Multilayer Graphene Nanoribbons. *Phys. Lett. A* **2012**, *376*, 525–528. [CrossRef]
19. Ansari, R.; Motevalli, B.; Montazeri, A.; Ajori, S. Fracture Analysis of Monolayer Graphene Sheets with Double Vacancy Defects via MD Simulation. *Solid State Commun.* **2011**, *151*, 1141–1146. [CrossRef]
20. Lee, S.H.; Lee, D.H.; Lee, W.J.; Kim, S.O. Tailored Assembly of Carbon Nanotubes and Graphene. *Adv. Funct. Mater.* **2011**, *21*, 1338–1354. [CrossRef]
21. Prezhdo, O.V.; Kamat, P.V.; Schatz, G.C. Virtual Issue: Graphene and Functionalized Graphene. *J. Phys. Chem. C* **2011**, *115*, 3195–3197. [CrossRef]
22. Wu, Y.H.; Yu, T.; Shen, Z.X. Two-Dimensional Carbon Nanostructures: Fundamental Properties, Synthesis, Characterization, and Potential Applications. *J. Appl. Phys.* **2010**, *108*, 071301. [CrossRef]
23. Vadukumpully, S.; Paul, J.; Valiyaveettil, S. Cationic Surfactant Mediated Exfoliation of Graphite into Graphene Flakes. *Carbon* **2009**, *47*, 3288–3294. [CrossRef]
24. Chae, H.K.; Siberio-Pérez, D.Y.; Kim, J.; Go, Y.; Eddaoudi, M.; Matzger, A.J.; O'Keeffe, M.; Yaghi, O.M. A route to high surface area, porosity and inclusion of large molecules in crystals. *Nature* **2004**, *427*, 523–527. [CrossRef]
25. Liu, Y.; Xie, B.; Zhang, Z.; Zheng, Q.; Xu, Z. Mechanical Properties of Graphene Papers. *J. Mech. Phys. Solids* **2012**, *60*, 591–605. [CrossRef]
26. Meng, Z.; Wu, D.; Wang, L.; Zhu, H.; Li, Q. Carbon Nanotube Glycol Nanofluids: Photo-Thermal Properties, Thermal Conductivities and Rheological Behavior. *Particuology* **2012**, *10*, 614–618. [CrossRef]
27. Halelfadl, S.; Estellé, P.; Aladag, B.; Doner, N.; Maré, T. Viscosity of Carbon Nanotubes Water-Based Nanofluids: Influence of Concentration and Temperature. *Int. J. Therm. Sci.* **2013**, *71*, 111–117. [CrossRef]
28. Mahbubul, I.; Khan, M.M.A.; Ibrahim, N.I.; Ali, H.M.; Al-Sulaiman, F.A.; Saidur, R. Carbon Nanotube Nanofluid in Enhancing the Efficiency of Evacuated Tube Solar Collector. *Renew. Energy* **2018**, *121*, 36–44. [CrossRef]
29. Chen, L.; Xie, H.; Li, Y.; Yu, W. Nanofluids Containing Carbon Nanotubes Treated by Mechanochemical Reaction. *Thermochim. Acta* **2008**, *477*, 21–24. [CrossRef]
30. Vryzas, Z.; Kelessidis, V.C. Nano Based Drilling Fluids: A Review. *Energies* **2017**, *10*, 540. [CrossRef]
31. Ismail, A.; Aftab, A.; Ibupoto, Z.; Zolkifile, N. The novel approach for the enhancement of rheological properties of water-based drilling fluids by using multi-walled carbon nanotube, nanosilica and glass beads. *J. Pet. Sci. Eng.* **2016**, *139*, 264–275. [CrossRef]
32. Hamze, S.; Cabaleiro, D.; Estellé, P. Graphene-based nanofluids: A comprehensive review about rheological behavior and dynamic viscosity. *J. Mol. Liq.* **2021**, *325*, 115207. [CrossRef]
33. Wang, Y.; Al-Saaidi, H.A.I.; Kong, M.; Alvarado, J.L. Thermophysical Performance of Graphene Based Aqueous Nanofluids. *Int. J. Heat Mass Transf.* **2018**, *119*, 408–417. [CrossRef]
34. Kole, M.; Dey, T.K. Investigation of thermal conductivity, viscosity, and electrical conductivity of graphene based nanofluids. *J. Appl. Phys.* **2013**, *113*, 084307. [CrossRef]
35. Moghaddam, M.B.; Goharshadi, E.K.; Entezari, M.H.; Nancarrow, P. Preparation, characterization, and rheological properties of graphene–glycerol nanofluids. *Chem. Eng. J.* **2013**, *231*, 365–372. [CrossRef]
36. Kazemi, I.; Sefid, M.; Afrand, M. A novel comparative experimental study on rheological behavior of mono & hybrid nanofluids concerned graphene and silica nano-powders: Characterization, stability and viscosity measurements. *Powder Technol.* **2020**. [CrossRef]
37. Ahammed, N.; Asirvatham, L.G.; Wongwises, S. Effect of volume concentration and temperature on viscosity and surface tension of graphene–water nanofluid for heat transfer applications. *J. Therm. Anal. Calorim.* **2016**, *123*, 1399–1409. [CrossRef]

38. Bashirnezhad, K.; Bazri, S.; Safaei, M.R.; Goodarzi, M.; Dahari, M.; Mahian, O.; Dalkılıça, A.S.; Wongwises, S. Viscosity of Nanofluids: A Review of Recent Experimental Studies. *Int. Commun. Heat Mass Transf.* **2016**, *73*, 114–123. [CrossRef]
39. Yang, L.; Xu, J.; Du, K.; Zhang, X. Recent Developments on Viscosity and Thermal Conductivity of Nanofluids. *Powder Technol.* **2017**, *317*, 348–369. [CrossRef]
40. Rasheed, A.; Khalid, M.; Rashmi, W.; Gupta, T.; Chan, A. Graphene Based Nanofluids and Nanolubricants – Review of Recent Developments. *Renew. Sustain. Energy Rev.* **2016**, *63*, 346–362. [CrossRef]
41. Sawicka, D.; Cieśliński, J.T.; Smolen, S. A Comparison of Empirical Correlations of Viscosityand Thermal Conductivity of Water-Ethylene Glycol-Al_2O_3 Nanofluids. *Nanomaterials* **2020**, *10*, 1487. [CrossRef]
42. Freddawati, R.W.; Amgad, A.A.; Kanji, Y.; Manaf, H.A. Seed/Catalyst-Free Growth of Gallium-Based Compound Materials on Graphene on Insulator by Electrochemical Deposition at Room Temperature. *Nanoscale Res. Lett.* **2015**, *10*, 233. [CrossRef]
43. Xin, W.; Long, Z. Green and facile production of high-quality graphene from graphite by the combination of hydroxyl radicals and electrical exfoliation in different electrolyte systems. *RSC Adv.* **2019**, *9*, 3693–3703. [CrossRef]
44. Cabaleiro, D.; Colla, L.; Barison, S.; Lugo, L.; Fedele, L.; Bobbo, S. Heat Transfer Capability of (Ethylene Glycol + Water)-Based Nanofluids Containing Graphene Nanoplatelets: Design and Thermophysical Profile. *Nanoscale Res. Lett.* **2017**, *12*, 53. [CrossRef] [PubMed]
45. Wang, S.; Wang, C.; Ji, X. Towards Understanding the Salt-Intercalation Exfoliation of Graphite into Graphene. *RSC Adv.* **2017**, *7*, 52252–52260. [CrossRef]
46. Nizar, A.; Godson, A.L.; Joel, T.; Raja, B.J.; Somchai, W. Measurement of Thermal Conductivity of Graphene–Water Nanofluid at below and above Ambient Temperatures. *Int. Commun. Heat Mass Transf.* **2016**, *70*, 66–74. [CrossRef]
47. Baby, T.T.; Ramaprabhu, S. Investigation of Thermal and Electrical Conductivity of Graphene Based Nanofluids. *J. Appl. Phys.* **2010**, *108*, 124308. [CrossRef]
48. Chen, H.; Ding, Y.; Lapkin, A. Rheological behaviour of nanofluids containing tube/rod-like nanoparticles. *Powder Technol.* **2009**, *194*, 132–141. [CrossRef]
49. Bird, R.B.; Stewart, W.E.; Lightfoot, E.N. *Transport Phenomena*; John Wiley and Sons, Inc.: Hoboken, NJ, USA, 2002.
50. Vogel, D.H. Das Temperaturabhaengigkeitsgesetz der Viskositaet von Fluessigkeiten. *Phys. Z.* **1921**, *22*, 645–646.
51. Vallejo, J.P.; Żyła, G.; Fernández-Seara, J.; Lugo, L. Influence of Six Carbon-Based Nanomaterials on the Rheological Properties of Nanofluids. *Nanomaterials* **2019**, *9*, 146. [CrossRef]
52. Phan-Thien, N. *Understanding Viscoelasticity*; Springer-Verlag: Berlin/Heidelberg, Germany, 2013. [CrossRef]
53. Kowalska, M.; Krztoń-Maziopa, A.; Babut, M.; Mitrosz, P. Rheological and physical analysis of oil-water emulsion based on enzymatic structured fat. *Rheol. Acta* **2020**, *59*, 717–726. [CrossRef]
54. Rosti, M.E.; Picano, F.; Brandt, L. chapter Numerical Approaches to Complex Fluids. In *Flowing Matter*; Springer: Berlin, Germany, 2019; pp. 1–35. [CrossRef]
55. Einstein, A. Eine neue Bestimmung der Moleküldimensionen. *Ann. Der Phys.* **1906**, *19*, 289–306. [CrossRef]
56. Brinkman, H. The viscosity of concentrated suspensions and solutions. *J. Chem. Phys.* **1952**, *20*, 571–581. [CrossRef]
57. Batchelor, G.K.; Green, J.T. The determination of the bulk stress in a suspension of spherical particles to order c^2. *J. Fluid Mech.* **1972**, *56*, 401–427. [CrossRef]
58. Krieger, I.M.; Dougherty, T.J. A mechanism for non-newtonian flow in suspensions of rigid spheres. *Trans. Soc. Rheol.* **1959**, *3*, 137–152. [CrossRef]
59. Vallejo, J.; Gomez-Barreiro, S.; Cabaleiro, D.; Gracia-Fernandez, C.; Fernández-Seara, J.; Lugo, L. Flow behaviour of suspensions of functionalized graphene nanoplatelets in propylene glycol–water mixtures. *Int. Commun. Heat Mass Transf.* **2018**, *91*, 150–157. [CrossRef]
60. Cabaleiro, D.; Estellé, P.; Navas, H.; Desforges, A.; Vigolo, B. Dynamic Viscosity and Surface Tension of Stable Graphene Oxide and Reduced Graphene Oxide Aqueous Nanofluids. *J. Nanofluids* **2018**, *7*, 1081–1088. [CrossRef]
61. Loulijat, H.; Koumina, A.; Zerradi, H. The Effect of the Thermal Vibration of Graphene Nanosheets on Viscosity of Nanofluid Liquid Argon Containing Graphene Nanosheets. *J. Mol. Liq.* **2019**, *276*, 936–946. [CrossRef]
62. Zhang, H.; Wang, S.; Lin, Y.; Feng, M.; Wu, Q. Stability, thermal conductivity, and rheological properties of controlled reduced graphene oxide dispersed nanofluids. *Appl. Therm. Eng.* **2017**, *119*, 132–139. [CrossRef]
63. Mehrali, M.; Sadeghinezhad, E.; Rosen, M.A.; Akhiani, A.R.; Latibari, S.T.; Mehrali, M.; Metselaar, H.S.C. Experimental investigation of thermophysical properties, entropy generation and convective heat transfer for a nitrogen-doped graphene nanofluid in a laminar flow regime. *Adv. Powder Technol.* **2016**, *27*, 717–727. [CrossRef]
64. Sadeghinezhad, E.; Mehrali, M.; Latibari, S.T.; Mehrali, M.; Kazi, S.; Oon, S.; Metselaar, H. Experimental Investigation of Convective Heat Transfer Using Graphene Nanoplatelet Based Nanofluids under Turbulent Flow Conditions. *Ind. Eng. Chem. Res.* **2014**, *53*, 12455–12465. [CrossRef]
65. Ding, Y.; Alias, H.; Wen, D.; Williams, R.A. Heat transfer of aqueous suspensions of carbon nanotubes. *Int. J. Heat Mass Transf.* **2006**, *49*, 240–250. [CrossRef]

Article

Effect of Shear Flow on Nanoparticles Migration near Liquid Interfaces

Ali Daher [1,2,*], Amine Ammar [1], Abbas Hijazi [2] and Lazhar Benyahia [3]

[1] LAMPA, ENSAM Angers, 2 Boulevard du Ronceray, BP 93525, CEDEX 01, 49035 Angers, France; amine.ammar@ensam.eu
[2] Faculty of Sciences1, MPLAB, Lebanese University, Beirut 6573, Lebanon; abhijaz@ul.edu.lb
[3] Institut des Molécules et Matériaux du Mans (IMMM), UMR 6283 CNRS—Le Mans Université, Avenue Olivier Messiaen, CEDEX 09, 72085 Le Mans, France; lazhar.benyahia@univ-lemans.fr
* Correspondence: a.phys12@live.com

Citation: Daher, A.; Ammar, A.; Hijazi, A.; Benyahia, L. Effect of Shear Flow on Nanoparticles Migration near Liquid Interfaces. *Entropy* **2021**, *23*, 1143. https://doi.org/10.3390/e23091143

Academic Editor: Mikhail Sheremet

Received: 28 July 2021
Accepted: 25 August 2021
Published: 31 August 2021

Publisher's Note: MDPI stays neutral with regard to jurisdictional claims in published maps and institutional affiliations.

Copyright: © 2021 by the authors. Licensee MDPI, Basel, Switzerland. This article is an open access article distributed under the terms and conditions of the Creative Commons Attribution (CC BY) license (https://creativecommons.org/licenses/by/4.0/).

Abstract: The effect of shear flow on spherical nanoparticles (NPs) migration near a liquid–liquid interface is studied by numerical simulation. We have implemented a compact model through which we use the diffuse interface method for modeling the two fluids and the molecular dynamics method for the simulation of the motion of NPs. Two different cases regarding the state of the two fluids when introducing the NPs are investigated. First, we introduce the NPs randomly into the medium of the two immiscible liquids that are already separated, and the interface is formed between them. For this case, it is shown that before applying any shear flow, 30% of NPs are driven to the interface under the effect of the drag force resulting from the composition gradient between the two fluids at the interface. However, this percentage is increased to reach 66% under the effect of shear defined by a Péclet number $Pe = 0.316$. In this study, different shear rates are investigated in addition to different shearing times, and we show that both factors have a crucial effect regarding the migration of the NPs toward the interfacial region. In particular, a small shear rate applied for a long time will have approximately the same effect as a greater shear rate applied for a shorter time. In the second studied case, we introduce the NPs into the mixture of two fluids that are already mixed and before phase separation so that the NPs are introduced into the homogenous medium of the two fluids. For this case, we show that in the absence of shear, almost all NPs migrate to the interface during phase separation, whereas shearing has a negative result, mainly because it affects the phase separation.

Keywords: liquid–liquid interface; shear rate; nanoparticles; diffuse interface; phase field method; molecular dynamics; numerical simulation

1. Introduction

In recent years, using nanoparticles (NPs) in industrial and medical markets has grown significantly. They have been of immense significance in different branches of science and engineering. This is basically due to their unique properties, such as augmented reactivity and special optical properties, which make them very suitable for products and applications in tissue engineering, composite technology, enhanced oil recovery and drug delivery [1,2]. In addition, NPs arise in nanoparticle-armored fluid droplets [3] and phase-arrested gels [4]. It is well known that studying biological processes on the nanoscale level is an essential point behind the development of nanotechnology [5].

The assembly of NPs at the liquid–liquid interface is essential in the preparation and stabilization of conventional emulsions, which have wide applications in petroleum, cosmetics, food, and biological transferring [6,7]. Modeling the dynamics of NPs at liquid–liquid interfaces has a crucial role in developing static and dynamic flow models that help in drug delivery and understanding the biological and physical phenomena inside the cells of the body [8].

In addition, the dynamics of NPs in non-aqueous media, such as ionic liquids (ILs), was reported in many studies. It was shown that the design and preparation of the nanomaterials are well planned and executed, using ILs to produce tunable and functional fluid ILs-based nanomaterials, and it also was reported that ILs help to synthesize nanomaterials with various functionalized surfaces [9].

Furthermore, the non-extensivity of entropy was investigated for different sizes of colloidal Ag nanoparticles (NPs), and it was shown that the subextensivity of entropy occurs for colloidal Ag NPs. In the small size of colloidal Ag NPs and at low temperature, nonextensivity is important [10]. Taherkhani et al. used classical molecular dynamics (MD) simulations to investigate the radial distribution, glass transition, ionic transfer number, and electrical conductivity of the ionic liquid 1-ethyl-3-methylimidazolium hexafluorophosphate [EMIM][PF6] ionic liquid encapsulated in carbon nanotube (CNT), and they also studied the effect of nitrogen as a doping element in CNT on these properties of [EMIM][PF6] by MD simulation. It was shown that in the presence of nitrogen, ion transfer uses a hydrogen bonding mechanism, while in its absence, ion transfer uses a diffusion mechanism in which the cation has a significant effect on the ion transfer and electrical conductivity [11].

The behavior of NPs is strongly affected by the surrounding environmental factors and thus, external effects will modify their dynamic properties. Many researchers have studied the effects of external fields on the aggregation of NPs. The effect of shear on nanoparticle dispersion in polymer melts was investigated by Karla et al. [12], and it was shown that shear significantly slows down the aggregation of NPs and such an effect is strongly dependent on the polymer chain length and shear rate. In addition, Karla et al. [13] studied how the NPs disperse in a block copolymer system under shear flow; they found that shear can have a pronounced effect on the location of NPs in block copolymers and that it can be used as a parameter to control nano-composite self-assembly. In addition, Minh D Vo et al. [14] used dissipative particle dynamics (DPD) methods to study the effect of shear and particle shape on the physical adsorption of a polymer on carbon nanoparticles; they found that there are three possible states of the polymer adsorption on carbon nanoparticles (adsorbed, shear affected, and separated states) depending on the value of the shear rate. Besides that, the effects of shear stress on the intracellular uptake of NPs in a biomimetic microfluidic system were investigated by Kang et al. [15], and they showed that for the case of cationic NPs, as the magnitude of the shear stress increases, the intracellular uptake of such NPs maximizes at a certain value of shear stress and then decreases gradually, which ensures that the shear stress has a crucial role in various nanoparticles and drug delivery systems.

Plater et al. investigated experimentally the effect of viscosity ration of polypropylene/Poly-e caprolactone blends on the localization of carbon black aggregates. The authors reported that the particles were dragged to the viscous phase, even when the particles were located initially into the more fluid phase, although they preferred to locate in the latter [16].

Becu et al. succeeded in visualizing a single armored droplet with nanoparticles undergoing a shear flow in another Newtonian medium. The results showed a continuous but clear slowdown of the droplet relaxation after successive strain jumps. This effect is related to the densification of the droplet interface by NPs when deformed [17].

In the current study, we focus on how shear force affects the migration of NPs to the liquid–liquid interface, which will help to understand how NPs behave under standard industrial processing conditions. The goal of this study is to implement a compact model for the simulation of the shear effect on the migration of spherical NPs near a liquid–liquid interface. Our work is based on the phase field method (PFM) for the fluids modeling, using the diffuse interface model, and molecular dynamics (MD) for modeling the motion of the nanoparticles, through which we superimpose the discrete model of NPs (using MD) on the continuum model of fluids (using PFM), which is a new idea in numerical modeling that we discussed briefly in our previous paper [18].

The content of this paper is as follows. In Section 2, we give details of the models and methods that we used in numerical simulation. We describe the diffuse interface model and give a brief description of the numerical implementation and time discretization. In Section 3, we discuss the numerical results for the migration of nanoparticles at the liquid–liquid interface. We investigate the effect of shearing on the migration of nanoparticles at the liquid–liquid interface, and we compare the results for different shear rates and different shearing times. Finally, a conclusion follows in Section 4.

2. Materials and Methods

2.1. Particle–Particle and Fluid–Particle Interactions

In this section, the discrete dynamic of particles is described. Molecular dynamics is the method of choice when one wants to study the dynamical properties of a system in full atomic detail, provided that the properties are observable within the time scale accessible to simulations. Time scale is one of the two main limitations of the method as will be discussed later. Molecular dynamics simulations are also useful when the system cannot be studied by the experimental methods mentioned above, for example, when the protein cannot be crystallized or is too big or insoluble to be studied by NMR.

To calculate the dynamics of the system, that is, the position of each atom as a function of time, Newton's classical equation of motion is solved iteratively for each atom.

Each NP is considered a rigid spherical body whose velocity (v_i) and position (x_i) are updated by using Newton's equation of motion, which relates the applied forces with the particle's acceleration (a_i) according to the following equation:

$$F_i(t) = m_i \frac{d^2 x_i}{dt^2} = m_i a_i(t) \tag{1}$$

where F_i is the applied force on the particle 'i', and m_i is the mass of the particle. The applied forces can be classified into particle–particle interaction forces and external forces due to the fluid (in our case, we consider drag forces, Brownian forces and shear effects).

The force on each atom is the negative of the derivative of the potential energy with respect to the position of the atom:

$$f_{j-i} = -\nabla V(r) \tag{2}$$

If the potential energy of the system is known, then, given the coordinates of a starting structure and a set of velocities, the force acting on each atom can be calculated and a new set of coordinates generated, from which new forces can be calculated. Repetition of the procedure will generate a trajectory corresponding to the evolution of the system in time. The accuracy of the simulations is directly related to the potential energy function used to describe the interactions between particles. In molecular dynamics, a classical potential energy function is used that is defined as a function of the coordinates of each of the atoms. In macroscopic systems, the fraction of the particles near the wall is negligible, whereas in the MD simulations, this fraction is more significant, and the surface effect is of great importance. In order to reduce the surface effect and conserve the number of particles in the simulation box, periodic boundary conditions must be used so that when a particle enters or leaves the simulation region, an image particle leaves or enters this region.

The position and velocity of the NPs (xi and vi respectively) at each new time are calculated, using the velocity Verlet algorithm time integration scheme:

$$x(t+\delta) = x(t) + v(t)\delta + \frac{1}{2} a(t)\delta^2 \tag{3}$$

$$v(t+\delta) = v(t) + \frac{1}{2}[a(t) + a(t+\delta)]\delta^2 \tag{4}$$

where δ indicates the time increments for the molecular dynamics (MD).

In our model, the particle–particle interaction is calculated, using the truncated Lennard–Jones (LJ_{trunc}) potential; this potential is used in many numerical studies in order to model particle–particle interactions. See [19] for example.

$$V^{LJ}_{ij_{trunc}}(r) = \begin{cases} V_{LJ}(r_{ij}) - V_{LJ}(r_c) & \text{for } r < r_c \\ 0 & \text{for } r > r_c \end{cases} \quad (5)$$

With

$$V_{LJ}(r) = 4\epsilon \left[\left(\frac{\sigma}{r}\right)^{12} - \left(\frac{\sigma}{r}\right)^6 \right] \quad (6)$$

where r_c is the cut-off distance, which is taken to be 3σ, ϵ is the depth of the potential well, σ denotes the equilibrium distance, and r is center to center separation between two particles.

Thus, the interaction force acting on the $i-th$ particle induced by the $j-th$ particle is given by the following:

$$f_{j \to i} = -\nabla V_{LJ_{trunc}}(r) \, n_{j \to i} = -\frac{\partial V_{LJ_{trunc}}(r_{ij})}{\partial r_{ij}} n_{j \to i} \quad (7)$$

where $r_{ij} = r_j - r_i$ is the separation distance between two nanoparticles, i and j correspond to different NPs indices, and $n_{j \to i}$ represents the unit vector that point from x_j to x_i.

So, for N particles, the force acting on each particle is formed by the individual interactions with all the neighboring particles:

$$F_i = \sum_{j \neq i}^{N} f_{j \to i} \quad (8)$$

The external forces acting on the NPs are due the surrounding fluids and thus, each NP is affected by the Brownian force F_r and the drag force F_d given by the following formulas:

$$F_r(t) = \sqrt{2 D \Delta t} \, \chi \quad (9)$$

where Δt is the time step in the numerical model, D is the diffusion coefficient and χ is a normal random number whose average is zero and variance is one.

The drag effects are considered by the following:

$$F_d = \varphi \left(v_f - v_p \right) \quad (10)$$

where φ is the drag coefficient, v_p is the particle's velocity and v_f is the fluid's velocity resulting from the Navier–Stokes equation involving the concentration gradient (see [20] for details):

$$v_f = div(\nabla c \otimes \nabla c) \quad (11)$$

where $\otimes$ corresponds to the tensor product. The definition of c is introduced in the next section. It denotes the phase field scalar used to describe the mixture.

2.2. Diffuse Interface Model

In the diffuse interface model, the convective Cahn–Hilliard equation is given by the following:

$$\frac{\partial c}{\partial t} + u.\nabla c - \nabla.j_A = 0 \quad (12)$$

The diffusive flux is given by $j_A = M\nabla\mu$, where M denotes the mobility (scalar in the case of isotropic separation mixture, and tensor in the case of non-isotropic separation). μ is the chemical potential obtained from the variational derivative of the free energy with respect to the mass fraction c.

The Cahn–Hilliard theory [21] assumes that the driving force for diffusion is the gradient of the chemical potential, and thus, the above equation is generally written as follows:

$$\frac{\partial c}{\partial t} + u.\nabla c = \nabla.(M\nabla \mu) \tag{13}$$

The free energy is a double-sink function that implies that the only stable equilibrium values of c are +1 or −1. We use the classical symmetric form of ψ, which is used in most cases. However, other forms of non-symmetric potential can be used. The expression of free energy is assumed given by the following:

$$f(c, \nabla c) = -\frac{1}{2}\alpha c^2 + \frac{1}{4}\beta c^4 + \frac{1}{2}\varepsilon|\nabla c|^2 \tag{14}$$

where ε is the gradient energy parameter and α and β are positive constants. Thus, we have the following:

$$\mu = \frac{\delta f}{\delta c} = -\alpha c + \beta c^3 - \varepsilon \nabla^2 c \tag{15}$$

For this study, we apply shear forces. In order for the expression of this force to be compatible with our framework of periodic study, we consider that the shear force is taken to be periodic along the y-axis and defined by the following:

$$f_{shear} = \frac{\sin 2 * \pi * y}{l} \tag{16}$$

This is illustrated in Figure 1.

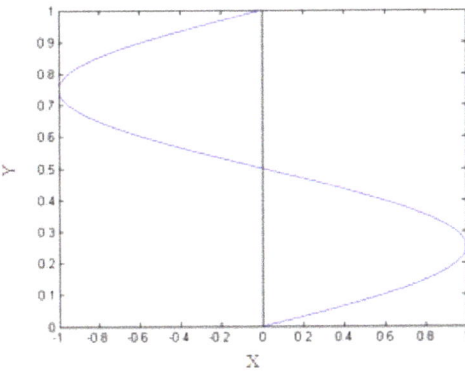

Figure 1. Periodic shear force along the y-axis.

The effect of shearing is introduced into the momentum equation represented by the Navier–Stokes equation (NS) with a phase field-dependent surface force as follows [22]:

$$\rho\left(\frac{\partial u}{\partial t} + (u.\nabla)u\right) = -\rho\nabla g + \nabla.\eta\left(\nabla u + \nabla u^T\right) + \rho\,\mu\nabla c + f_{shear} \tag{17}$$

We also consider the continuity equation for an incompressible fluid as follows:

$$\nabla.u = 0 \tag{18}$$

In the above equations, the dynamic viscosity of the fluid is denoted by η, u is the velocity field and g is the Gibbs free energy given by the following:

$g = f + p/\rho$, where p is the pressure and ρ is the mass density. The superscript T stands for the transpose operator.

2.3. Scaling the Equations

In order to simplify the equations and minimize the effects of round-off errors, it is preferable to use a set of dimensionless parameters. So, we scale the governing equations by defining U_c and L_c as the characteristic velocity and characteristic length, respectively, $T_c = \frac{L_c}{U_c} = \sigma.\sqrt{\frac{m}{\epsilon}}$ as the characteristic time and ε_c as the characteristic energy. The characteristic length of the phase field scale is related to that of the molecular dynamics as $L_c = 100\,\sigma$ (fixed ratio chosen for our studies).

We introduce $r^* = r/\sigma$, $\varepsilon^* = \varepsilon/\varepsilon_c$, $c^* = \frac{c}{c_B}$, $u^* = \frac{u}{U_c}$, $t^* = \frac{tU_c}{L_c}$, $\mu^* = \frac{\mu\varepsilon^2}{\epsilon c_B}$, $\eta^* = \frac{\eta}{\eta_c}$ as the normalized viscosity.

In addition, the dimensionless drag coefficient is defined as follows:

$$\varphi^* = \frac{\varphi\,\sigma}{\sqrt{\varepsilon_c\,m}}$$

The Péclet number is defined as the product of a shear rate by a characteristic time as follows:

$$Pe = \dot{\gamma}.\,T_c = \dot{\gamma}.\,\sigma.\sqrt{\frac{m}{\epsilon}}$$

$\xi = \sqrt{\frac{\epsilon}{\alpha}}$ Is the interfacial thickness and $c_B = \sqrt{\frac{\alpha}{\beta}}$ is the bulk concentration, which represents the mean field equilibrium value. Dropping the asterisk notations, we obtain the following:

$$\frac{dc}{dt} = -\boldsymbol{u}.\nabla c + \frac{1}{Pe}\nabla^2\mu \tag{19}$$

$$\mu = c^3 - c - C\nabla^2 c \tag{20}$$

$$\nabla.\boldsymbol{u} = 0 \tag{21}$$

$$\nabla g - \nabla.\eta\left(\nabla\boldsymbol{u} + \nabla\boldsymbol{u}^T\right) = \frac{1}{C.C_a}\mu\nabla c + f_{shear} \tag{22}$$

$$V_{LJ}(r) = 4\epsilon\left[\left(\frac{1}{r}\right)^{12} - \left(\frac{1}{r}\right)^{6}\right] \tag{23}$$

From this potential, we can get the force to be the following:

$$F_{LJ} = -\nabla V_{LJ}(r) \tag{24}$$

$$F_d = \varphi * \left(v_f - v_p\right) \tag{25}$$

We obtain the following set of dimensionless groups: the Cahn number C and the Capillary number C_a, defined respectively as follows:

$$C = \frac{\xi}{L_c}\;;\; C_a = \frac{\xi\eta U_c}{\rho\epsilon c_B{}^2};$$

2.4. Numerical Implementation

To model the dynamics of the NPs, we use the molecular dynamics (MD) method with periodic boundary conditions implemented in order to conserve the number of NPs in the simulation box.

In addition, we use the phase field method in order to model the two fluids and the formation of the interface between them. In this method, the concentration is defined as in Equation (19).

To find the weak form, we multiply by a weighting function c^* and integrate over the whole fluid domain Ω to obtain the following:

$$\int c^* \frac{dc}{dt} d\Omega + \int c^* u.(\nabla c) d\Omega - \int \frac{1}{P_e} c^* \nabla^2 \mu \, d\Omega = 0 \qquad (26)$$

where $c^* = \underline{c}^{*T} \underline{N}$ and N is defined as $[\ N_1 \ \ N_2 \ \ N_3 \ \ N_4\ ]$.

In the above equation $N_1, \ldots N_4$ are the quadratic interpolation functions of the 4-node quadrilateral element:

$$\frac{dc}{dt} = \underline{N}^T \underline{\dot{c}} \qquad (27)$$

The finite element interpolation for the gradient of the concentration is described in terms of the linear combination of the shape function derivatives, given in matrix form, by the following:

$$\nabla c = \mathbf{B} \, \underline{c} \qquad (28)$$

$$\mathbf{B} = \begin{bmatrix} N_{1,x} & N_{2,x} & N_{3,x} & N_{4,x} \\ N_{1,y} & N_{2,y} & N_{3,y} & N_{4,y} \end{bmatrix}$$

Solving Equation (23), we get:

$$\underline{c}^{*T} \int \underline{N} \, \underline{N}^T \, d\Omega \, \underline{\dot{c}} + \underline{c}^{*T} \int \underline{N} u . \mathbf{B} \, d\Omega \, \underline{c} - \underline{c}^{*T} \frac{1}{P_e} \int N \nabla^2 \mu \, d\Omega = 0 \qquad (29)$$

This integration allows obtaining a linear system that has to be solved at each time step, which can be solved using a semi-implicit or explicit time integration scheme:

$$M \, \underline{\dot{c}} + G \, \underline{c} + F(\underline{c}) = 0 \qquad (30)$$

Similarly, in order to solve the velocity equation, we can write the following:

$$\int u^* \nabla^2 u \, d\Omega + \int u^* A \mu \nabla c \, d\Omega = 0 \qquad (31)$$

$$\text{where } u^* = \underline{u}^{*T} \underline{N} \qquad (32)$$

$$\text{and } \nabla^2 u = K \, \underline{u} \qquad (33)$$

Solving Equation (28), we obtain the following:

$$\underline{u}^{*T} \int N \, K \, d\Omega \, \underline{u} + \underline{u}^{*T} \int N \, A \, \mu \, B \, d\Omega \, \underline{c} = 0 \qquad (34)$$

This integration gives the following linear system to be solved at each time step:

$$T \, \underline{u} + H \, \underline{c} = 0 \qquad (35)$$

The position and velocity of the NPs (x_i and v_i respectively) at each new time are updated, using the velocity Verlet integration scheme.

3. Results and Discussion

3.1. Fluids Separation and Interface Formation

The system is composed of two liquids that are normally immiscible. Due to some external effects (temperature for example), these two liquids may be mixed to form one thermodynamic phase. So, we can start the study from an initial state where the two fluids are totally mixed as shown in Figure 2 (left figure).

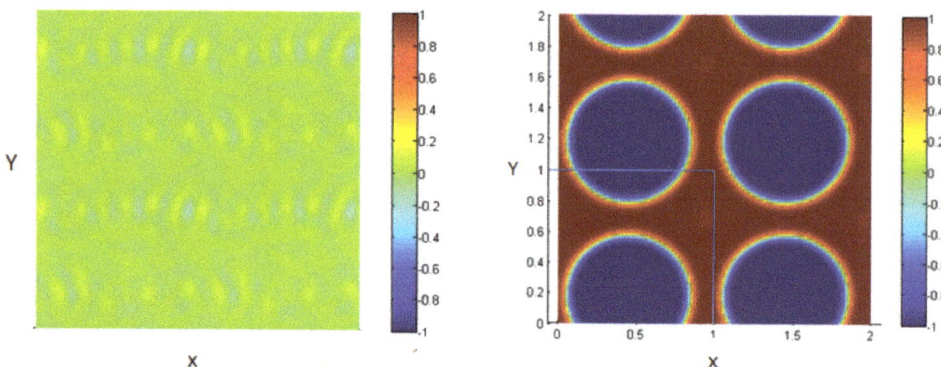

Figure 2. Mixture of two immiscible liquids, from the homogenous state (**left**) to the equilibrium state after the phase separation (**right**).

These figures represent a 2D plot of the simulation box representing the two fluids.

The side bar in Figure 2 represents the evolution of the concentration between the two fluids in order to form the interface. The concentration is taken to vary from -1 in the first fluid (blue) to 1 in the second one (red). In our model, the area fractions of the two phases are taken, as shown in Table 1.

Table 1. Variation of concentrations through the medium of the two fluids.

Concentration "C"	Liquid 1	Liquid 2
-1	100%	0%
1	0%	100%
0	50%	50%

As time goes on, and since the two liquids are normally immiscible, the molecules of each fluid immediately start to cluster together into microscopic clusters throughout the liquid. These clusters then rapidly grow and coalesce until we obtain an equilibrium state in which the two fluids are totally separated, and the interface is formed between them, as shown in Figure 2 (right).

The blue medium represents the first liquid, and the red represents the second liquid; the concentration is varying, according to the diffuse interface model. Note that our simulation box is bounded between 0 and 1 along the two axes as shown by the limiting lines in the figure, but we represent a periodic repetition of this box in Figure 2 (right) for clarity; this is done in all 2D figures throughout the paper.

3.2. Introduction of Nanoparticles after the Separation of the Two Fluids and the Formation of the Interface

After the two fluids are separated and the interface is formed between them, we introduce N nanoparticles randomly into the system.

We consider two cases regarding the concentration of NPs in the medium (total area of the NPs relative to the area of the medium of two fluids). The first case is 0.06 (200 NPs), and the second is 0.3 (1000 NPs).

3.2.1. Low Concentration of Nanoparticles

Consider 200 NPs (concentration of NPs 0.06) distributed randomly within the two fluids as shown in Figure 3.

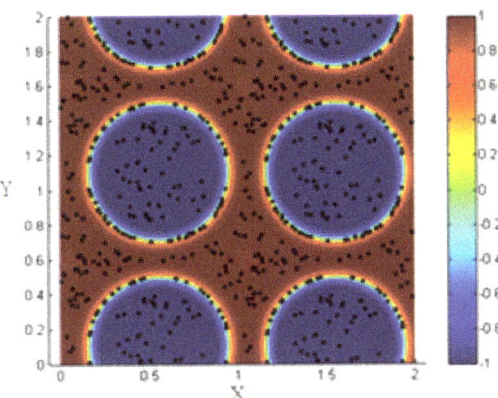

Figure 3. Random distribution of 200 NPs within the two fluids after phase separation.

- Neglect particle–particle interactions

For the first moment, let us neglect the interaction between the NPs via LJ potential and study the migration of NPs toward the interface. In this part, we examine the behavior of the NPs once they are introduced into the mixture of the two fluids that is already separated and the interface is formed between them. The main goal here is to study whether the external shear effect can improve the migration of NPs to the interface or not. Different cases regarding the shear effect are simulated. We start with the case involving no shear, and then we increase the Péclet number progressively; in each case, the percentage of NPs migrating to the interface is plotted.

1. No shear case.

Once introduced into the medium of the two fluids, the NPs near the interface are affected by the drag force given in Equation (7), resulting from the concentration gradient between the two fluids at the interface. These NPs are driven to the interfacial region, whereas the NPs far from the interface are not affected by this force, and thus, only 37% of the NPs migrate to the interface as shown in Figure 4.

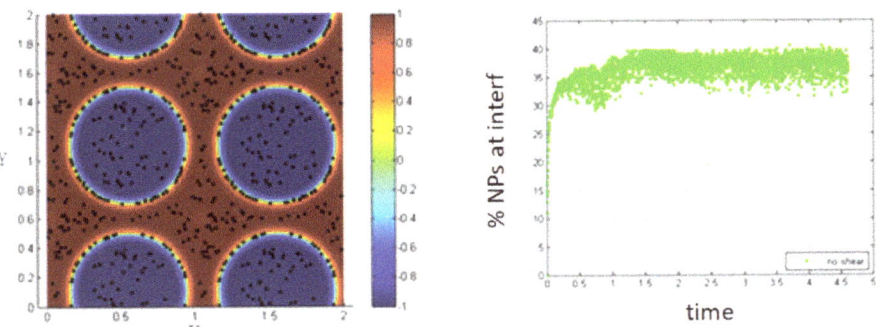

Figure 4. NPs near the interface migrate to the interfacial region (**left**); percentage of NPs belonging to the interfacial region in the absence of shear (**right**).

In the simulation, time is defined as a dimensionless parameter (normalized time) $t^* = \frac{tU_c}{L_c}$ as shown in Section 2.3 (scaling the equations).

The percentage of NPs belonging to the interface is calculated by finding the number of NPs belonging to the interfacial region relative to the total number of NPs introduced into the medium.

In this work, the interface is defined as the region of high concentration gradient, and we are able to track the position of the NPs and determine those that reach the interfacial thickness by calculating the concentration gradient at every position. By this way, we are able to determine whether each NP reaches the interface or not. It is important to note that the noise in Figure 4 (right) and all the coming figures are mainly due to the effect of the Brownian force introduced in the system. Although the effect of the other forces dominates the effect of the Brownian force, there are still some effects as seen by the percentage of error caused due to the Brownian force After the steady state is reached after introducing NPs, a shear is applied for a certain duration (T-shear), defined relative to the characteristic time. Thus, under the effect of the drag force, NPs near the interface are adsorbed. We consider different cases for which we quantify the evolution of the NPs percentage belonging to the interfacial region.

2. Simulation with shear rate = 0.4; Pe = 0.008, T-shear = 0.3.

In Figure 5, MD corresponds to molecular dynamics simulation, and it represents the separation of the two fluids in the absence of shear (before applying shear). This figure shows that the percentage of NPs migrating to the interface in the absence of shear is about 37%, and this percentage is not significantly affected in the case of low shear (low Péclet number Pe = 0.008).

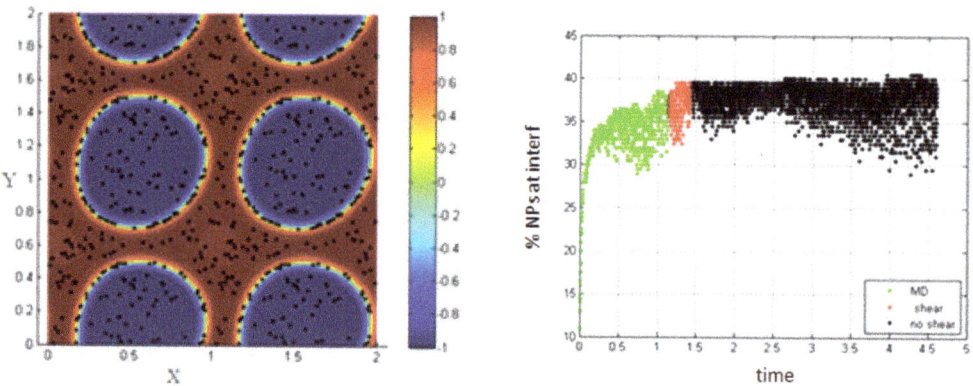

Figure 5. Shape deformation of the two fluids under the effect of shear (**left**); percentage of NPs belonging to the interfacial region in the regions, before applying shear (green), during applying shear (red), and after stopping shearing (black), Pe = 0.008 (**right**).

For this case, we find that a low Péclet number does not have a noticed effect on the percentage of NPs belonging to the interface.

3. Simulation with shear rate = 7.7; Pe = 0.154, T-shear = 0.3.

Increasing the shear rate and thus, increasing the Péclet number to 0.154, the percentage of NPs belonging to the interfacial region increases under the effect of shear, from 37% to about 60% as shown in Figure 6.

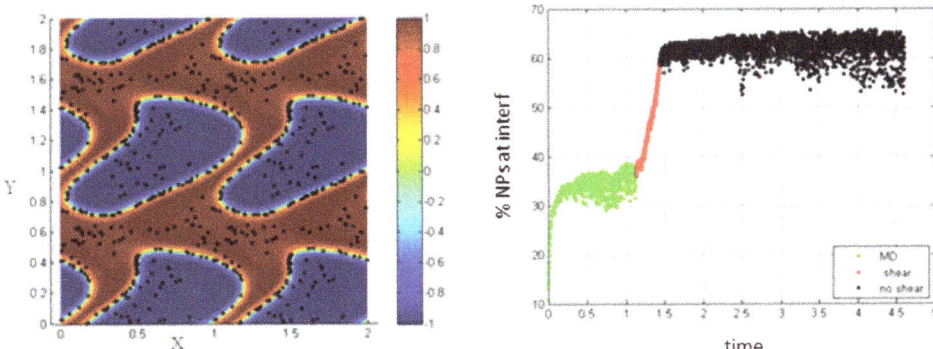

Figure 6. Percentage of NPs belonging to the interfacial region in the regions, before applying shear (green), during applying shear (red) and after stopping shearing (black), $Pe = 0.154$.

4. Simulation with shear rate =15.8; $Pe = 0.316$, T-shear = 0.3.

In Figure 7, it is clear that increasing the shear rate and thus, increasing the Péclet number, enhances the migration of the NPs toward the interface to reach about 62% at the end of the shear; also, this percentage increases a little bit to reach 70% after stopping shearing.

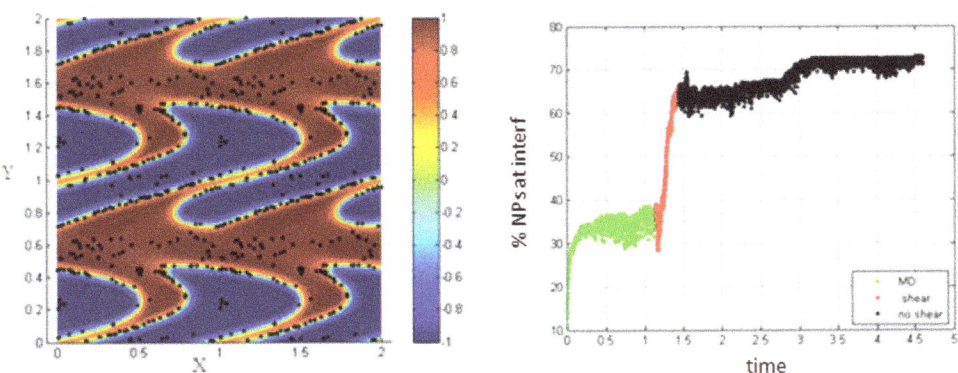

Figure 7. Percentage of NPs belonging to the interfacial region in the regions, before applying shear (green), during applying shear (red) and after stopping shearing (black), $Pe = 0.316$.

This is mainly due to the fact that as the shear rate increases, more interfaces are formed in the medium (i.e., the length of the interface is increasing) and thus, the percentage of NPs belonging to the interfacial region increases. After stopping shearing, the interfaces tend to reach an equilibrium state, and they reach the separation phase. In this case, the NPs that are still near the interface are attracted to the interfacial region, due to the concentration gradient so that the percentage increases a little bit to reach 70%.

- Including particle–particle interactions.

In this part, we consider the same situation discussed above (concentration of NPs 0.06), but this time we take into account the particle–particle interactions.

Once introduced, the NPs near the interface are driven by the hydrodynamic drag force to the interfacial region, and all the NPs that are close to each other interact by the particle–particle interaction force, so the NPs form clusters around each other as shown in Figure 8.

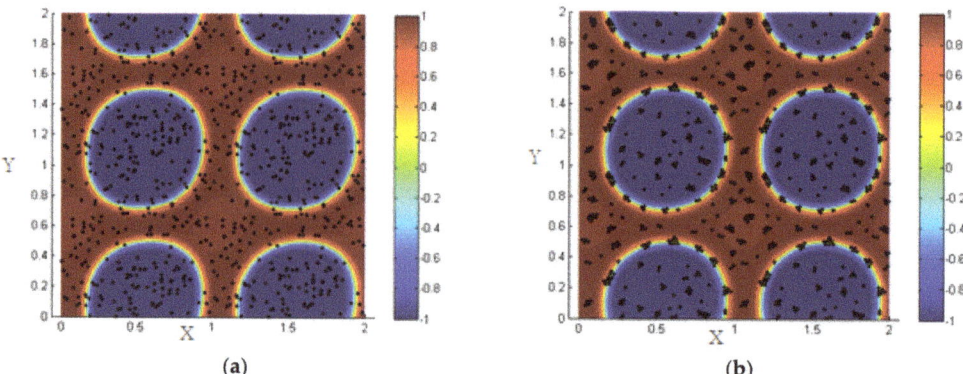

Figure 8. (**a**) Random distribution of the NPs within the two fluids; (**b**) formation of clusters under the effect of particle–particle interactions.

In this part, we also study the effect of different shear rates.

1. No shear case.

Evaluating the percentage of NPs that are within the interfacial thickness, before applying any shear on the system, shows that this percentage is about 26% as shown in Figure 9. This value is smaller than that found in the case of neglecting the Lennard–Jones potential. This is mainly because the formation of clusters prevents the clustering NPs from reaching the interface. In reality, they agglomerate around the ones that are in the interfacial thickness so that the cluster is attached to the interface by some of the NPs forming it.

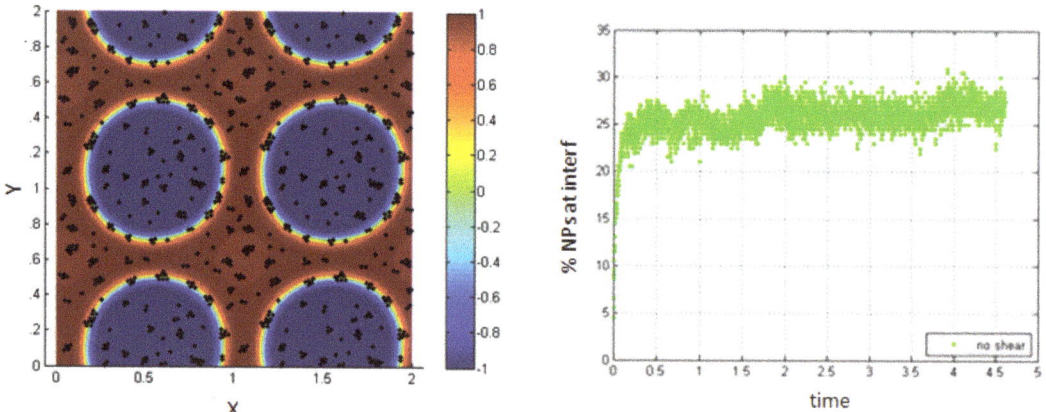

Figure 9. NPs near the interface migrate to the interfacial region (**left**); percentage of NPs belonging to the interfacial region in the regions, before applying shear (**right**).

The distribution of the NPs at the interface and the accumulation of the others over them are shown in Figure 8 above. In this case, the NPs that are attached to those at the interface are seen in the figure to belong to the first fluid (red) or the blue one (blue) and thus, they are not considered to belong to the interface. They are blocked by the ones that reached the interface before.

Now, we apply shear after the NPs near the interface are adsorbed, and we investigate the effect of shear on the migration of NPs to the interfacial region.

2. Simulation with $Pe = 0.008$; T-shear = 0.3.

As for the case without the Lennard–Jones potential, small shear rates do not have an important influence on the system. Thus, the percentage of NPs belonging to the interfacial region is not significantly modified, compared to the case of no shear as shown in Figure 10.

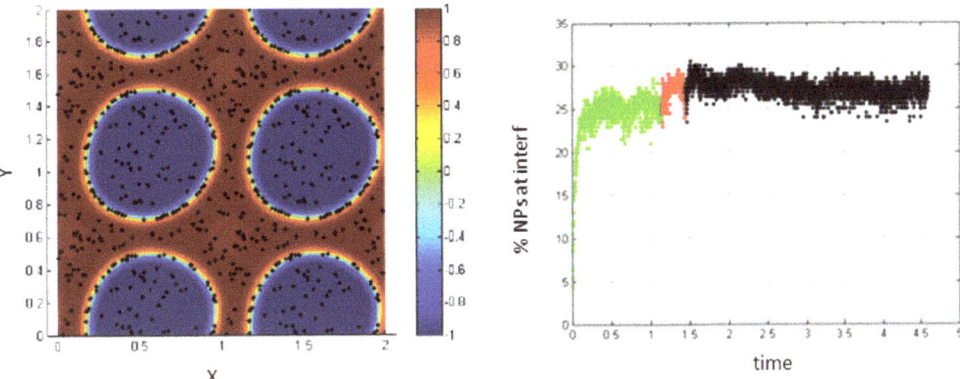

Figure 10. NPs near the interface migrate to the interfacial region (**left**); percentage of NPs belonging to the interfacial region in the regions, before applying shear (green), during applying shear (red) and after stopping shearing (black), $Pe = 0.008$ (**right**).

3. Simulation with shear rate =15.8; $Pe = 0.316$; T-shear = 0.3.

As the shear rate increases, more interfaces are formed in the medium (i.e., the length of the interface is increasing) and thus, the percentage of NPs belonging to the interfacial region increases, as shown in Figure 11.

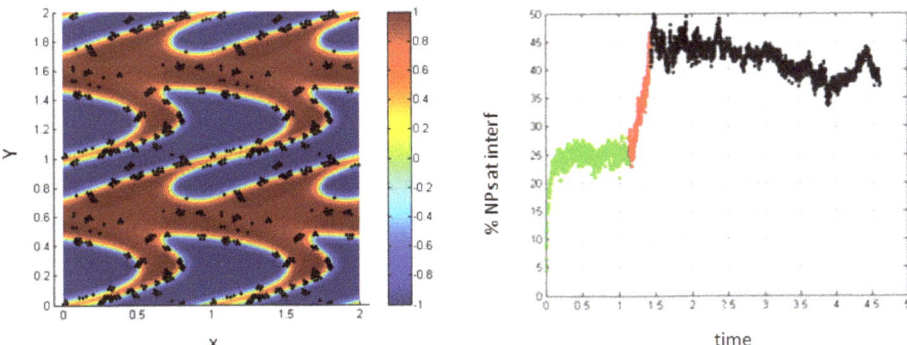

Figure 11. Percentage of NPs belonging to the interfacial region in the regions, before applying shear (green), during applying shear (red) and after stopping shearing (black), $Pe = 0.316$.

One can notice that the percentage in these cases differs from that in the cases discussed above, where we neglected the Lennard–Jones potential. This is mainly due to the fact that in the case when the particle–particle interactions are taken into account, the NPs accumulate around each other and form clusters at the interface.

It is noticed in the literature that embedding particles at the liquid interfaces may, for example, lead to increased stability of biphasic systems, such as Pickering's emulsions [23], or lead to a double percolation morphology followed by electrical conductivity as with immiscible polymer blends [24]. For the latter case, several studies have focused on the competition or synergy between thermodynamics and hydrodynamics that is inerrant to mixing processes [25–28] on the particle localization. However, the observation of the particle adsorption dynamics remains rarely considered. For example, Keal et al. were able to demonstrate, by confocal microscopy tracking, the adsorption by natural diffusion of colloidal particles at a liquid–liquid interface of very low interfacial tension. To date, we are not aware of any experimental work describing the adsorption dynamics of nanoparticles at the liquid–liquid interfaces under flow.

The blue boxes (marked area) in Figure 12 show that the two interfaces come so close, and the particles accumulate between them.

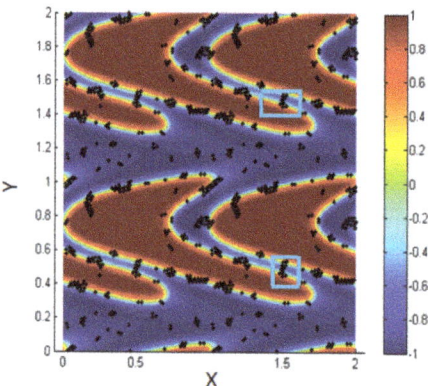

Figure 12. Some clusters are attached between two adjacent interfaces, $Pe = 0.316$.

3.2.2. Simulation with High Concentration of NPs

As for the previous concentration, we take into account two cases. First, we neglect the Lennard–Jones (L.J.) potential and then we include it in the second case.

- Simulation of neglecting the L.J.potential.

For the initial state, consider 1000 NPs (concentration of NPs 0.3) randomly distributed within the two fluids, as shown in Figure 13a.

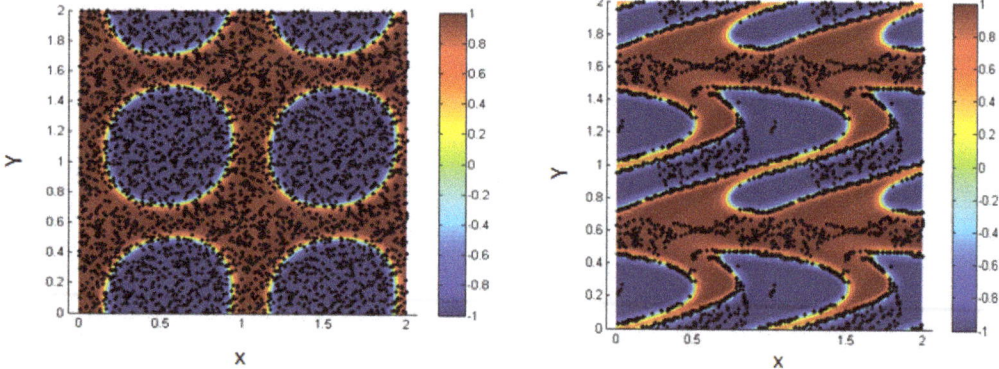

Figure 13. Random distribution of the NPs within the two fluids (**left**); more NPs migrate to the interface under the effect of shear, $Pe = 0.316$, T-shear = 0.3 (**right**).

First, neglecting the effect of the L.J. potential, we discuss the effect of different shear rates applied for the same duration (T-shear), and in addition, we study the effect of different shearing durations for the same shear rates. i.e., for the same Péclet numbers.

By increasing the Péclet number to 0.316, we find that about 66% of the NPs are driven to the interfacial thickness for shear duration (T-shear = 0.3). Thus, we again ensure that shearing enhances the migration of the NPs toward the liquid–liquid interface.

- Simulation including the L.J. potential.

In this part, we include the effect of the Lennard–Jones potential, for the case of 1000 NPs within the simulation box. We fix $Pe = 0.3160$ and T-shear = 0.3, and we examine different cases through which we vary certain parameters related to the L.J. potential.

1. $\epsilon = 1, \sigma = 0.02$

In Figure 14, it is clear that by increasing the concentration of NPs within the simulation box, each NP has an interaction with all the NPs close to it. Due to the high concentration of NPs, all of them are connected with each other and thus, the L.J. force is strong enough to prevent the shear from affecting the motion and the location of the NPs.

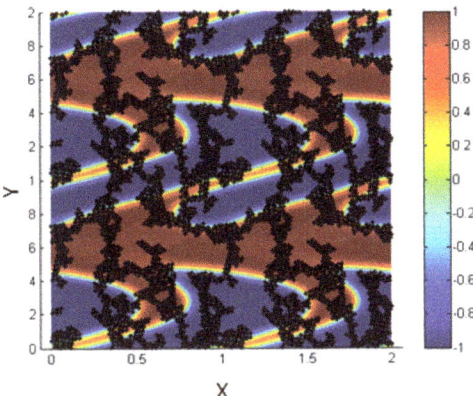

Figure 14. Particle–particle interactions are stronger than the effect of shear.

2. Simulation with $\epsilon = 0.1, \sigma = 0.02$

In experimental studies, it is possible to control and vary the value of the potential between each NP and its neighbors and thus, it is possible to modify the particle–particle interactions in order to match some industrial needs [29,30].

In order to study the effect of shearing in such cases, we minimize the value of the potential well's depth, ϵ, in the definition of the L.J. potential.

The results are shown in Figure 15, where we find that in such cases, shearing affects the motion of the NPs and let them migrate toward the interfacial region. In addition, there are some NPs connected between two adjacent interfaces, and this matches the experimental results reported by Becu and Benyahia [14].

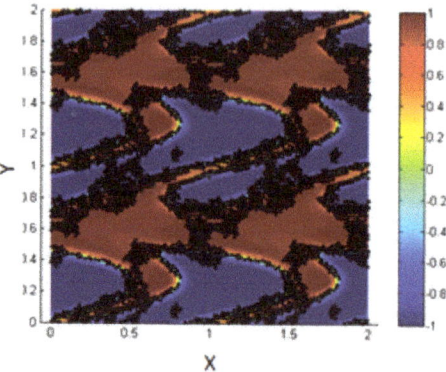

Figure 15. Migration of the NPs clusters to the interface under the effect of shear.

It is clear from the obtained results that by controlling the value of the potential between neighboring NPs and under the effect of shear, we are able to let the NPs form a layer on the interface, even when the concentration of the NPs in the fluids is high.

3.3. Introduce the NPs Randomly into the Mixture of the Two Fluids before Phase Separation

In this part, we study the case of NPs introduced into the medium of the homogenous mixture of the two fluids before phase separation and thus, before the formation of the interface as shown in Figure 16.

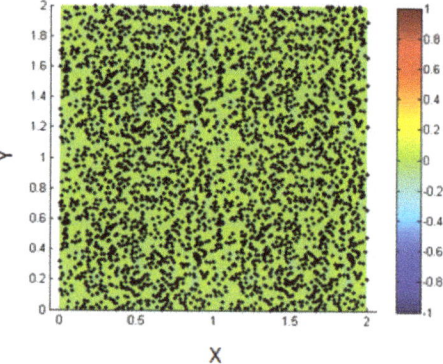

Figure 16. Initial state of NPs randomly distributed within the mixed fluids.

3. Simulation with no shear.

As time goes on, and since the two fluids are immiscible, they will start to separate, but this time the NPs are within the fluids and thus, their motion is affected by the phase separation.

In the absence of shear, it is clear from Figure 17 that almost all the NPs (100%) are driven to the interfacial region after a time of 2.2 relative to the characteristic time. The percentage increases progressively during the phase separation so that the NPs are affected by the force resulting from the concentration gradient throughout all the phase separation time.

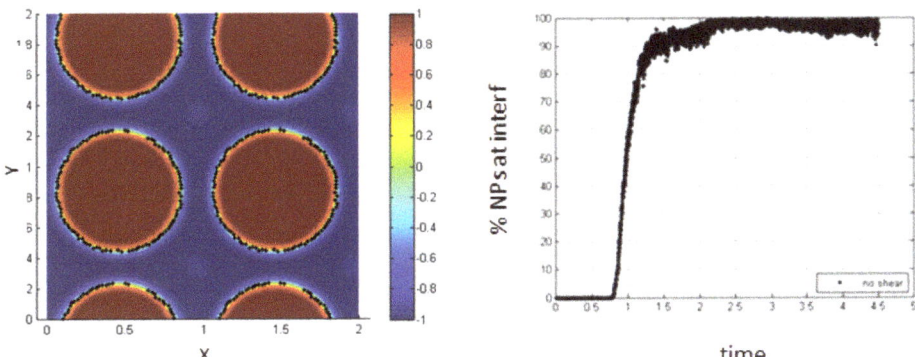

Figure 17. Percentage of NPs belonging to the interfacial region, in the absence of shear.

Comparing these results to the case when the NPs are introduced after phase separation has occurred, it is clear that the time of NPs introduction is important.

Since the percentage of NPs adsorbing at the interface is less when the phase separation is already occurred, this seems to indicate that the potential of adsorption is higher in this case. However, when we start from the homogenous state, the potential barrier is less, and thus, more particles can be adsorbed at the interface.

In order to quantify the effect of shear on the migration of the NPs to the interface in this case, we present the results obtained for different shear rates applied to the system just at the beginning of the simulation before the phase separation starts.

4. Simulation with shear rate = 0.4; $Pe = 0.008$

As for the previous cases, a small shear rate does not have an important effect on the obtained results, compared to the case of no shear as shown in Figure 18.

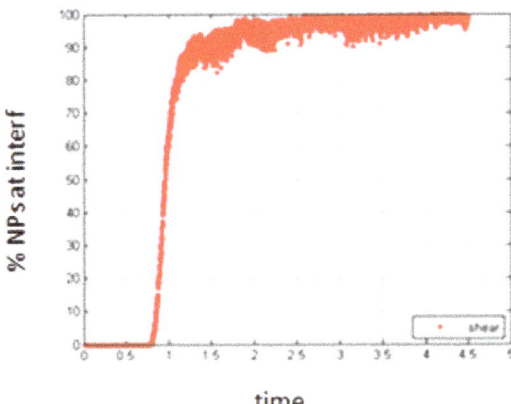

Figure 18. Percentage of NPs belonging to the interfacial region, in the presence of small shear.

5. Simulation with shear rate 7.7; $Pe = 0.154$.
6. Simulation with shear rate 15.8, $Pe = 0.316$

Increasing the shear rate will have a negative result with respect to the migration of NPs towards the interface. As seen in Figures 19 and 20, the percentage of NPs belonging to the interface is smaller compared to the case of no shear for all time durations.

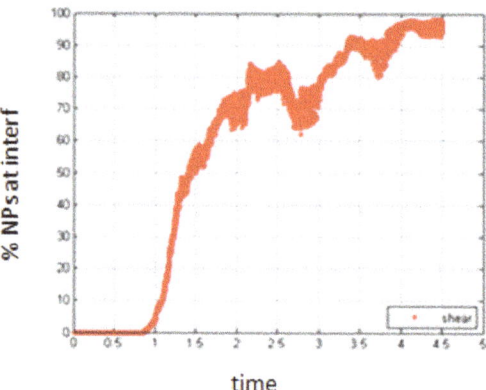

Figure 19. Percentage of NPs belonging to the interfacial region in the presence of shear, $Pe = 0.154$.

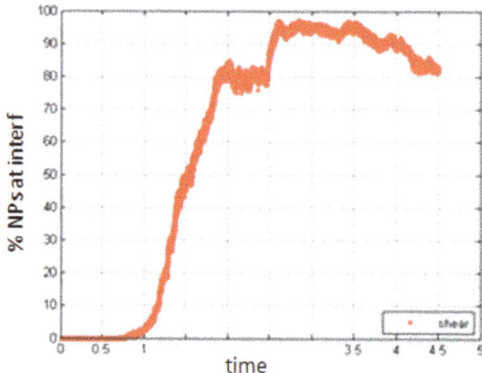

Figure 20. Percentage of NPs belonging to the interfacial region in the presence of shear, $Pe = 0.316$.

So as seen in the figures above, we find that in the case when we introduce the NPs into the medium of the two mixed fluids before phase separation takes place, more NPs will migrate to the interface in the absence of shear.

4. Conclusions

In this study, we have investigated the effect of shear force on the migration of nanoparticles toward the interface of two immiscible liquids.

We have implemented a numerical simulation through which we modeled the two fluids using the diffuse interface model. Two cases were investigated. The first one is to introduce the NPs randomly into the medium of two fluids that have been separated and the interface was formed between them. The second case is to introduce the NPs randomly into the mixture of the two fluids before phase separation takes place. For the first case, if we leave the fluids without any external effect, we find that only a small percentage, not greater than 30% of the NPs will migrate towards the interface. These NPs are the ones that are close to the interface when we introduce them randomly into the medium. Their migration toward the interface is mainly due to the effect of the drag force related to the concentration gradient in the interfacial region.

We have found that inducing a shear onto the system enhances the percentage of NPs belonging to the interface. It has been shown that there are two factors affecting the percentage of NPs migrating towards the interface, the value of the shear rate modified using the Péclet number, Pe and the shear duration T-shear. Introducing a shear defined by $Pe = 0.154$ for a duration of 0.63 drives 66% of the NPs to the interface compared to 30 % for the case of no shear effect. The same result was obtained using $Pe = 0.316$ for duration of 0.3. This ensures that both factors play a crucial role regarding the migration of NPs towards the interface.

In addition, we have discussed the effect of including or neglecting the particle-particle interaction using the Lennard–Jones potential and we have found that for low concentration of NPs, some differences appear regarding the percentage of NPs within the interface. This is mainly because due to particle–particle interactions NPs close to each other will form clusters and accumulate around each other, so that then NPs first attached to the interface will attract the close ones to accumulate around them so that we find clusters of NPs attached to the interface. In addition, for the same case we find that some clusters will be attached between two close interfaces which match with some experimental observations [14]. On the other hand, we have found that for the case of high concentration of NPs there are several observations depending on the value of the L.J. potential well. For $\epsilon = 1$ the particle–particle interaction will be strong enough and will compete the effect of shear and thus the NPs will not be driven to the interface. However, for $\epsilon = 0.1$ the NPs will be driven to the interface, where they form a layer at the interfacial region, with some of them are connected with other NPs on adjacent interfaces. From the experimental point of view, it is well known that we can modify the value of the particle–particle potential interactions in order to match some industrial needs and thus the NPs can be driven under the effect of shear to the interface even though when their concentration in the fluids is high. So for this case, we have shown that shearing is a key factor in enhancing the migration of the NPs towards the interface.

On the other hand, we have found that NPs introduced randomly into the mixture of the two fluids before phase separation takes place will migrate faster and in a higher percentage to the interface in the absence of shear.

It is clear that the time of NPs introduction is important. Since the percentage of NPs adsorbing at the interface is less when the phase separation already occurred, seems to indicate that the potential of adsorption is higher in the case. Whereas, when we start from the homogenous state, the potential barrier is less, thus, more particles can be adsorbed at the interface.

These results help us to understand how NPs behave under standard industrial processing conditions. We obtain more information regarding new methods for synthesizing nanomaterial films; in addition, these results also help to understand the behavior of small NPs within the cells of organisms where they are greatly affected by the flow rate of the surrounding fluids.

Author Contributions: Data curation, A.D.; Formal analysis, A.D., A.H. and L.B.; Funding acquisition, A.H.; Investigation, A.D.; Methodology, A.D.; Project administration, A.H., A.A., Software, A.D. and A.A.; Supervision, A.A. and A.H.; Validation, A.D.; Writing—original draft, A.D.; Writing—review & editing, A.D. All authors have read and agreed to the published version of the manuscript.

Funding: This work was supported by grant (2020) from the Lebanese University.

Conflicts of Interest: The authors declare no conflict of interest.

References

1. Bhushan, B. *Handbook of Nanotechnology*; Springer: Berlin/Heidelberg, Germany, 2006; pp. 17–45.
2. ShamsiJazeyi, H.; Miller, C.A.; Wong, M.S.; Tour, J.M.; Verduzco, R. Polymer-coated nanoparticles for enhanced oil recovery. *J. Appl. Polym.* **2014**, *131*, 40576. [CrossRef]
3. Cauvin, S.; Clover, P.J.; Bon, S.A.F. Pickering stabilized miniemulsion polymerization: Preparation of clay armored latexes. *Macromolecules* **2005**, *38*, 7887–7889. [CrossRef]
4. Clegg, P.S. Fluid-bicontinuous gels stabilized by interfacial colloids: Low and high molecular weight fluids. *J. Phys. Condens. Matter* **2008**, *20*, 113101. [CrossRef] [PubMed]
5. Whitesides, G.M. The 'right' size in Nanobiotechnology. *Nat. Biotechnol.* **2003**, *21*, 1161–1165. [CrossRef]

6. Saleh, N.; Sarbu, T.; Sirk, K.; Lowry, G.V.; Matyjaszewski, K.; Tilton, R.D. Oil-in-water emulsions stabilized by highly charged polyelectrolyte-grafted silica nanoparticles. *Langmuir* **2005**, *21*, 9873. [CrossRef] [PubMed]
7. Horozov, T.S.; Aveyard, R.; Clint, J.H.; Neumann, B. Particle zips: Vertical emulsion films with particle monolayers at their surfaces. *Langmuir* **2005**, *21*, 2330. [CrossRef]
8. Grabinski, C.; Sharma, M.; Maurer, E.; Sulentic, C.; Mohan Sankaran, R.; Hussain, S. The effect of shear flow on nanoparticle agglomeration and deposition in in vitro dynamic flow models. *Nanotoxicology* **2015**, *10*, 1–10. [CrossRef]
9. Arumugam, V.; Redhi, G.; Gengan, R. The application of ionic liquids in nanotechnology. *Fundam. Nanoparticles* **2018**, 371–400. [CrossRef]
10. Taherkhani, F.; Minofar, B. Static and dynamical properties of colloidal silver nanoparticles in [EMim][PF6] ionic liquid. *Ion. Liq.* **2019**, 129–144. [CrossRef]
11. Taherkhani, F.; Minofar, B. Effect of Nitrogen Doping on Glass Transition and Electrical Conductivity of [EMIM][PF6] Ionic Liquid Encapsulated in a Zigzag Carbon Nanotube. *J. Phys. Chem. C* **2017**, *121*, 15493–15508. [CrossRef]
12. Kalra, V.; Escobedo, F.; Jooa, Y.L. Effect of shear on nanoparticle dispersion in polymer melts: A coarse-grained molecular dynamics study. *J. Chem. Phys.* **2010**, *132*, 024901. [CrossRef] [PubMed]
13. Kalra, V.; Mendez, S.; Escobedo, F.; Jooj, Y.L. Coarse-grained molecular dynamics simulation on the placement of nanoparticles within symmetric diblock copolymers under shear flow. *J. Chem. Phys.* **2008**, *128*, 164909. [CrossRef]
14. Vo, M.D.; Papavassiliou, D.V. The effects of shear and particle shape on the physical adsorption of polyvinyl pyrrolidone on carbon nanoparticles. *Nanotechnology* **2016**, *27*, 325709. [CrossRef] [PubMed]
15. Kang, T.; Park, C.; Choi, J.S.; Cui, J.H.; Lee, B.J. Effects of shear stress on the cellular distribution of polystyrene nanoparticles in a biomimetic microfluidic system. *J. Drug Deliv. Sci. Technol.* **2016**, *31*, 130–136. [CrossRef]
16. Plattier, J.; Benyahia, L.; Dorget, M.; Niepceron, F.; Tassin, J.F. Viscosity-induced filler localisation in immiscible polymer blends. *Polymer* **2015**, *59*, 260–269. [CrossRef]
17. Becu, L.; Benyahia, L. Strain-Induced Droplet Retraction Memory in a Pickering Emulsion. *Langmuir* **2009**, *25*, 6678–6682. [CrossRef]
18. Daher, A.; Ammar, A.; Hijazi, A. Nanoparticles migration near liquid-liquid interfaces using diffuse interface model. *Eng. Comput.* **2019**, *36*, 1036–1054. [CrossRef]
19. Choi, Y.J.; Djilali, N. Direct numerical simulations of agglomeration of circular colloidal particles in two dimensional shear flow Phys. *Fluids* **2016**, *28*, 013304. [CrossRef]
20. Lowengrub, J.; Truskinovsky, L. Quasi-incompressible Cahn-Hilliard fluids and topological transitions. *Proc. R. Soc. A* **1998**, *454*, 2617–2654. [CrossRef]
21. Cahn, J.W.; Hilliard, J.E. Free energy of a nonuniform system. III. Nucleation in a two-component incompressible fluid. *J. Chem. Phys.* **1959**, *31*, 688–699. [CrossRef]
22. Choi, Y.J.; Anderson, P.D. Cahn–Hilliard modeling of particles suspended in two-phase flows. *Int. J. Numer. Methods Fluids* **2012**, *69*, 995–1015. [CrossRef]
23. Binks, B.P.; Horozov, T.S. (Eds.) *University of Hull. Colloidal Particles at Liquid Interfaces*; Cambridge University Press: Cambridge, UK, 2006; ISBN 9780511536670. [CrossRef]
24. Ibarra-Gomez, R.; Marquez, A.; Valle, L.; Rodriguez-Fernandez, O.S. Influence of the Blend Viscosity and Interface Energies on the Preferential Location of CB and Conductivity of BR/EPDM Blends. *Rubber Chem. Technol.* **2003**, *76*, 969–978. [CrossRef]
25. Elias, L.; Fenouillot, F.; Majeste, J.C.; Cassagnau, P. Morphology and rheology of immiscible polymer blends filled with silica nanoparticles. *Polymer* **2007**, *48*, 6029e40. [CrossRef]
26. Elias, L.; Fenouillot, F.; Majeste, J.C.; Alcouffe, P.; Cassagnau, P. Immiscible polymer blends stabilized with nano-silica particles: Rheology and effective interfacial tension. *Polymer* **2008**, *49*, 4378–4385. [CrossRef]
27. Elias, L.; Fenouillot, F.; Majeste, J.C.; Cassagnau, P. Morphology and Rheology of Polymer Blends Filled with Silica Nano Particles. *J Polym. Sci.Part B Polym. Phys.* **2008**, *46*, 1976–1983.
28. Keal, L.; Colosqui, C.E.; Tromp, R.H.; Monteux, C. Colloidal Particle Adsorption at Water-Water Interfaces with Ultralow Interfacial Tension. *Phys. Rev. Lett.* **2018**, *120*, 208003. [CrossRef] [PubMed]
29. Watson, K.J.; Zhu, J.; Nguyen, S.T.; Mirkin, C.A. Hybrid Nanoparticles with Block Copolymer Shell Structures. *J. Am. Chem. Soc.* **1999**, *121*, 462. [CrossRef]
30. Skaff, H.; Ilker, M.F.; Coughlin, E.B.; Emrick, T. Preparation of cadmium selenide-polyolefin composites from functional phosphine oxides and ruthenium-based metathesis. *J. Am. Chem. Soc.* **2002**, *124*, 5729–5733. [CrossRef]

Article

Octree Optimized Micrometric Fibrous Microstructure Generation for Domain Reconstruction and Flow Simulation

Nesrine Aissa [1,2,*], Louis Douteau [1], Emmanuelle Abisset-Chavanne [3], Hugues Digonnet [1], Patrice Laure [4,5] and Luisa Silva [1,*]

1. High Performance Computing Insitute, Ecole Centrale de Nantes, 1 rue de la Noe, 44300 Nantes, France; louis.douteau@ec-nantes.fr (L.D.); hugues.digonnet@ec-nantes.fr (H.D.)
2. GeM Institute, Ecole Centrale de Nantes, 1 rue de la Noe, 44300 Nantes, France
3. I2M Institute, CNRS, Arts et Métiers Paris Tech, University Bordeaux, Bordeaux INP, 33400 Talence, France; emmanuelle.abisset-chavanne@ensam.eu
4. Laboratoire J.-A. Dieudonné, CNRS UMR 6621, Université de Nice-Sophia Antipolis, 06000 Nice, France; patrice.laure@mines-paristech.fr
5. Cemef, UMR CNRS 7635, Mines ParisTech, 06904 Sophia Antipolis, France
* Correspondence: nesrine.aissa@ec-nantes.fr (N.A.); luisa.rocha-da-silva@ec-nantes.fr (L.S.)

Citation: Aissa, N.; Douteau, L.; Abisset-Chavanne, E.; Digonnet, H.; Laure, P.; Silva, L. Octree Optimized Micrometric Fibrous Microstructure Generation for Domain Reconstruction and Flow Simulation. *Entropy* **2021**, *23*, 1156. https://doi.org/10.3390/e23091156

Academic Editor: Mikhail Sheremet

Received: 29 July 2021
Accepted: 30 August 2021
Published: 2 September 2021

Publisher's Note: MDPI stays neutral with regard to jurisdictional claims in published maps and institutional affiliations.

Copyright: © 2021 by the authors. Licensee MDPI, Basel, Switzerland. This article is an open access article distributed under the terms and conditions of the Creative Commons Attribution (CC BY) license (https://creativecommons.org/licenses/by/4.0/).

Abstract: Over recent decades, tremendous advances in the field of scalable numerical tools and mesh immersion techniques have been achieved to improve numerical efficiency while preserving a good quality of the obtained results. In this context, an octree-optimized microstructure generation and domain reconstruction with adaptative meshing is presented and illustrated through a flow simulation example applied to permeability computation of micrometric fibrous materials. Thanks to the octree implementation, the numerous distance calculations in these processes are decreased, thus the computational complexity is reduced. Using the parallel environment of the ICI-tech library as a mesher and a solver, a large scale case study is performed. The study is applied to the computation of the full permeability tensor of a three-dimensional microstructure containing 10,000 fibers. The considered flow is a Stokes flow and it is solved with a stabilized finite element formulation and a monolithic approach.

Keywords: octree optimization; microstructure generation; domain reconstruction; flow simulation; permeability computing

1. Introduction

The properties and behavior of a discontinuous fiber-reinforced thermoplastic are induced by the mechanisms involved during the forming process. Modeling and numerical simulation have a major role in understanding and predicting these mechanisms, especially at the microscopic scale, which provides the most accurate results. Nevertheless, at this scale of computation, numerical simulations are generally expensive in terms of computing resources and time. Optimizing and evaluating the used algorithms is a constant challenge. One of the most expensive issues when using finite elements and immersed boundary approaches for discontinuous reinforced composites simulation is the computation of distances. Fibers generation, immersion, and reconstruction techniques particularly rely on these evaluations, as the distances between fibers must be regularly evaluated during microstructure generation and distances from each point of the computational mesh to the frontiers of the immersed elements have to be measured. However, without any optimization, whenever the number of points and fibers in a simulation rises, the cost of reconstruction increases dramatically. In order to make these techniques applicable in the context of composites materials, an optimization of the distance evaluation is required. A first idea is to implement distance computation algorithms that save computational time. Reducing the number of expensive functions or operations used to compute each distance is a key element, as well as defining properly the data types used to limit memory footprint.

This paper proposes a reduction in the number of distances to evaluate, which is performed using an octree.

The octree data structure [1] is a partition of a three-dimensional space built from recursive subdivisions into eight sub-domains. The sub-cubes obtained are hierarchically organized, which allows to easily reduce search time. Octree algorithms are widely used in various fields and their application range is significantly extensive, especially when positions must be accessed and manipulated. These applications include construction of a three-dimensional object model from a set of images [2] and simulation of displacement of free surface [3]. Octrees are broadly applied for collision detection algorithms in virtual reality, rigid bodies contacts, characters animation, or machining simulation, such as cutter-path generation for numerical control machines which require efficient collision detection routines [4–6]. Another significant example involving octree algorithm is the mesh generation procedure. Octree can be used to create meshes tied to geometrical objects [7], for adaptive mesh refinement (AMR), e.g., with structured grids in fluid dynamics [8], or combined with others techniques in advanced mesh generation processes [9].

In this paper, an octree-optimized microstructure generation and domain reconstruction with adaptative mesh is presented. An application of flow simulation through the reconstructed domains dealing with the identification of the full-component permeability tensor is conducted.

2. Microstructure Generation and Optimization Using Octree

The microstructure of a discontinuous fiber composite greatly affects its properties. For that, virtual numerical sample generation is crucial in order to carry out precise prediction simulations. However, a major difficulty to generate such a microstructure lies in the establishment of an optimized methodology that allows generating a very large number of fibers without interpenetration and with a minimum computation time and resources. In this work, a Random Sequential Adsorption RSA algorithm [10,11], widely used for rigid particles generation, is chosen.

A collection of N random unit orientations **P**, N homogeneously distributed mass center positions **X** and N lengths **L**, following a normal distribution law with mean length $<L>$ and standard deviation σ, are primarily created. The program begins with one initial fiber (i) randomly oriented with $\mathbf{P}_i$. Subsequently, another fiber (j) with a random orientation $\mathbf{P}_j$ is selected and then the system is checked for overlap. If the fiber (j) intersects a pre-existing fiber, it is repositioned by randomly changing orientation vector $\mathbf{P}_j$ while retaining the same position vector $\mathbf{X}_j$. The selection of a new $\mathbf{P}_j$ is repeated up to a maximum number of trials until the overlap condition is released. In this method, the generated geometry is periodic, so that any fiber cutting a boundary will be extended on the opposite one. This means that fibers close to surfaces can interact with the fibers of the near domains. Therefore, every new fiber to be placed is verified for interaction with all already pre-existing fibers and their 26 periodic images in the near domains.

Figure 1 presents an example of a generated microstructure with 1000 cylindrical fibers having a same diameter d, a mean aspect ratio $r = <L>/d = 20$ and a fiber volume fraction $V_f = 0.1$.

In the previously described algorithm, $N*27$ distances evaluations are required to generate the $N+1$-th fiber. Presuming that no intersection is detected, a minimum of $27*N*(N-1)/2$ distances has to be computed, thus leading to a N^2 complexity. However, this number can increase, as once an intersection occurs, new random positions and orientations must be generated for the fiber. This computational cost is acceptable when N remains small, but becomes unaffordable when N reaches the order of the millions of fibers. To limit the number of distances to evaluate, this paper proposes the use of an octree algorithm. This tree structure enables to browse rapidly across all the elements and to select them based on their position. Consequently, a selection of the closest elements can be performed, which allows measuring the distances to these elements only. The complexity is decreased and can reach $N \log(N)$ for an optimal problem. The next paragraphs describe the octree building procedure, while the use of the octree is explained in Section 3.1.

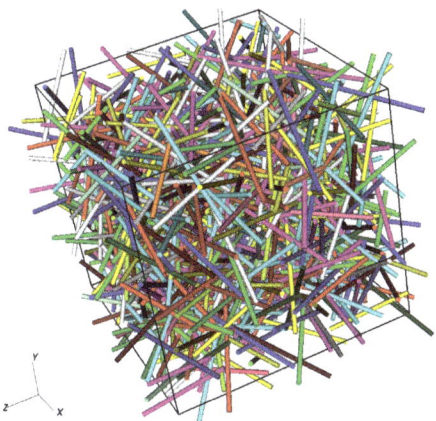

Figure 1. Example of a generated microstructure with 1000 fibers having a same diameter d, a mean aspect ratio $r = <L>/d = 20$ and a fiber volume fraction $V_f = 0.1$.

This data storage concept is a tree structure built recursively from a computational domain, in which elements, e.g., fibers, are dispersed. To clarify this paragraph, an analogy is performed between the computational domain and box bounding all the elements. In practice, there is a possibility for elements to be concentrated in a particular area of the computational domain. In that situation, the octree building procedure is processed in the interest region only, which does not cause any problem later on. The tree is built through refinement steps where the computational domain is divided in two along each dimension, thus generating subdomains (children). The name octree comes from the characterization of the tree in 3D, where 8 subdomains are generated by the division procedure (Figure 2).

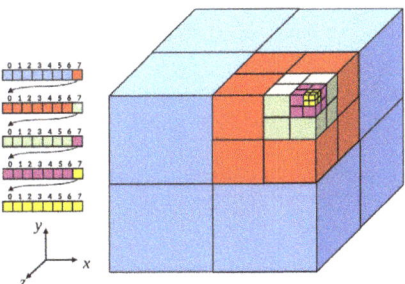

Figure 2. Illustration of the octree data structure: On the left is highlighted the refinement of a tree element into 8 new elements. The cube on the right presents the geometrical positions of the octree elements.

After refinement, the elements shall no more be contained in the initial computational domain, but are defined using pointers towards every child they intersect. This choice characterizes the octree class, which is composed of the dimensions of the computational domain and pointers to, either the elements contained inside it or the children generated. The corollary of this choice is that fibers can be duplicated if they intersect several children. After a refinement step, all the children are overlooked with emphasis on the number of elements it contains. If a subdomain remains empty, i.e., no elements intersect it, it is immediately deleted. If too many elements are found in this child, the refinement procedure is repeated in this particular subdomain. The recursiveness is applied in that way until: either an acceptable number of elements is obtained in the deepest subdomains (leaf), or the maximal depth of the octree is reached.

The repartition of elements into the children is handled using bounding boxes. Axis-Aligned Bounding Boxes **AABB** have been used, which offer different advantages. First of all, these boxes are very easy to determine, both computationally speaking and in terms of access to data. It also allows reducing the computational effort for the determination of the intersections, as the boxes are oriented along the same axes as the computational domain. Finally, this choice enabled to generalize the octree to very different usage, from fibers to, e.g., 3D facets used to define surface meshes. The drawback brought by these bounding boxes lies in the intersections, as an "ill-oriented" fiber may be duplicated in leaves it does not intersect, only because its bounding box does. In that case, we can implement Oriented Bounding Box **OBB** in future works to enclose fibers as tightly as possible. Another limitation occurs when very long elements (proportionally to the size of the computational domain) are present, as again the fibers may be highly duplicated. However, the following developments of this paper will show that octree usage remains appropriate for elements with a small length to width ratio.

This paragraph presents the octree generation on an example that features 14 fibers, with a maximal depth for the octree of 2 and 1 fiber allowed per leaf. The procedure is drawn in Figure 3.

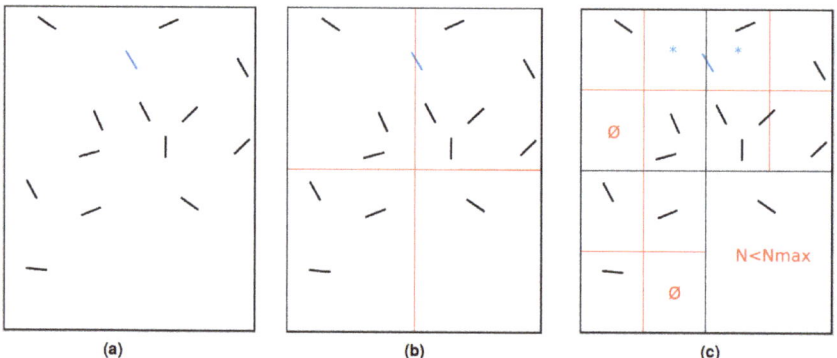

Figure 3. Octree generation example: (**a**) Fibers in computational domain. (**b**) Octree first level of refinement. (**c**) Octree second level of refinement.

The octree parameters mean that any subdomain containing more than 1 element needs to be refined, with a limit of only 2 levels. After the first step of the refinement, the fibers presented in Figure 3a are allocated to every subdomain their bounding box intersect. An interesting emphasis can be placed on the blue fiber (second "row" from the top, middle of the computational domain), which is duplicated into both of the two children on top of the initial computational domain in Figure 3. Consequently, after a second step of refinement this fiber can be found in two different octree leaves, the asterisked ones in Figure 3b. Figure 3c corresponds to the final octree as obtained with the parameters detailed previously. Even if the presence of only one fiber per leaf was authorized, subdomains containing more than one fiber can be found because of maximum refinement allowed. Note that the subdomains containing ∅ have been created by octree refinement, and immediately deleted as no fiber was allocated to it.

When adding a new fiber following the RSA algorithm, thanks to the implementation of the octree, the check for overlap will be carried out among a reduced number of fibers initially judged by the octree as potential candidates for collision. Fibers with which there is a possible collision are the fibers in the leaf or leaves to which the new fiber belongs and whose **AABB**s intersect. Figure 4 shows a schematic diagram of this method: it shows a leaf of an octree (large box black) to which we would like to add the red fiber and where the blue and green fibers already exist. Thus, a possibility of intersections can only occur with blue fibers. The green fibers will not be concerned because their **AABB**s do not intersect the **AABB** of the red fiber.

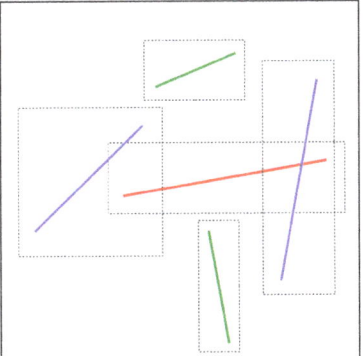

Figure 4. Illustration of the collision detection optimisation process: addition of the red fiber to an octree leaf, possibility of collision only with blues fibers whose AABBs intersect the AABB of the red one.

During this process, fibers are dynamically added to the octree. For that, two major conditions should be verified to update the octree after adding a new fiber:

- A new fiber must be always included in the global domain initially built for octree and, if it is not the case, it is necessary to destroy the octree and to reconstruct it;
- The size of a leaf should not exceed the defined maximal size and, if it is not the case, it is necessary to refine the octree.

To quantify the gain brought by the octree, we study the evolution of the CPU time, t, according to the number of generated fibers, N. For all the simulations, we consider $r = 20$, $V_f = 0.1$, and a maximum number of trials equal to 5000. The leaf maximal size is fixed to 100. Figure 5 shows a considerable gain on time which improves as the number of fibers becomes more important.

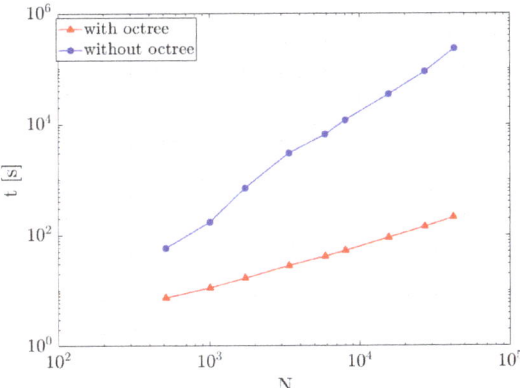

Figure 5. Microstructure generation time as a function of the number of fibers, for a case with microstructure having $r = 20$, $V_f = 0.1$, and a maximum number of trials equal to 5000.

3. Computational Domain Reconstruction

3.1. Mesh Immersion and Optimization Using the Octree

Mesh immersion is a technique enabling the representation of complex bodies using a single computational mesh. The main idea is to compute the distance from each point of the computational mesh to an object immersed, which can be represented by an analytical function, a mesh, or any set of data. The only constraint is the need to build an interior for the object, thus defining a frontier. This definition enables to establish a signed distance function α, as presented in Equation (1) for the immersion of a shape ω of the frontier $\Gamma = \partial \omega$ into a domain Ω. This interior can be concave or even split, as the mathematical evaluation of α does not have any prerequisite. However, the more complex ω will be, the more points in the computational mesh will be needed to represent it accurately.

$$\alpha = \bar{d}(x,\omega) = \begin{cases} d(x,\Gamma) & \text{if } x \in \omega \\ -d(x,\Gamma) & \text{if } x \notin \omega \end{cases}, x \in \Omega. \tag{1}$$

Once the signed distance function is defined, any computational point x has a signed-distance either positive or negative. The union of points with positive α defines the interior, and the inverse set gives the exterior. This formulation mathematically corresponds to using a Heaviside function as a level-set function, which gives 1 for α positive and 0 for α negative. However, this approach is not suitable for multiphase flows, as strong discontinuities are sources of instability when using Galerkin approximation for the resolution of the Navier–Stokes equations. To overcome this issue, a smoothed Heaviside function based on a width parameter ε has been defined and is presented in Equation (2).

$$H_\varepsilon(\alpha) = \frac{1}{2}\left(1 + \frac{u_\varepsilon(\alpha)}{\varepsilon}\right), \tag{2}$$

with

$$u_\varepsilon(\alpha) = \varepsilon \tanh\left(\frac{\alpha}{\varepsilon}\right). \tag{3}$$

This paradigm introduces a transition phase of a width of about 2ε which smooths the shifting between physical parameters of the two phases. The "blurred area" does not operate as a gray zone in terms of mesh immersion, as the norm and sign of the result given by H_ε in this region is depending on α. Compared to immersion results giving either 0 or 1 for a classical Heaviside function, a better capture of the interfaces can even be achieved. However, the quality of the reconstruction of ω remains highly dependent on the meshing of Ω. Fine meshes are needed around interfaces, and if the meshing of ω is complex, a high effort will be put in either mesh generation or distance evaluation.

This interdependency is addressed by coupling the immersion with a mesh adaptation procedure. An anisotropic mesh generated automatically concentrates its points around Γ, guaranteeing that an important portion of them will be located in the transition region highly impacted by H_ε. Further explanations about this procedure can be found in Section 3.2 and in [12]. Figure 6a presents the results of α for a circle of a radius R, and Figure 6b presents the results obtained for H_ε with $\varepsilon = R/100$. A slice of the computational mesh is also drawn, where the major part of the points are gathered in the interest zones (Figure 6c).

The level-set function is defined analytically from α, making the evaluation of α the major effort of the immersion procedure. If an analytical definition of α requires only one distance computation per point and does not need to be optimized, considering more complex representations generates computing complexity, e.g., when meshes or fibers set are immersed. Those cases use a set of elements to define ω or Γ, so the determination of the closest neighbor is not immediate. The performance of the immersion code then highly depends on the computational effort needed to evaluate a single distance, but also on the number of distances to compute before finding the closest element of ω. Without any optimization of the immersion procedure, the computation of α for a single point x and M fibers defining ω require M distance evaluations. Consequently, the immersion of M

fibers in Ω composed of N points forces the computation of $N \times M$ distances. When few fibers are immersed in small meshes, this cost is affordable. However, when 10,000 fibers are immersed, as proposed in the case of study of this paper, the number of computations is extremely high (assuming that N is quasi-linearly related to M), which is somewhere between not competitive and unrealizable computationally. The coupling of the mesh immersion procedure with an octree is a way to reduce the complexity. The construction of the octree was overlooked in Section 2, and its contribution to the reduction in computational costs is detailed in the next paragraphs.

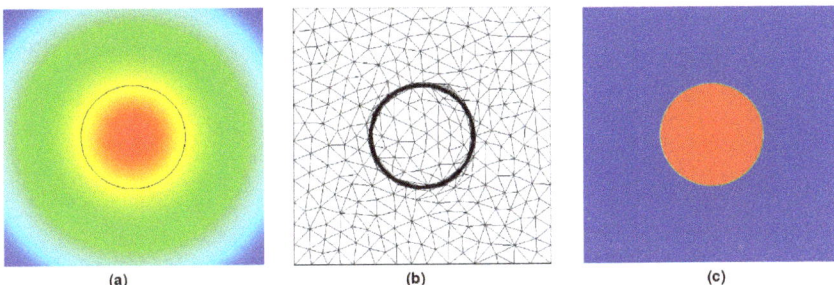

Figure 6. Immersion and mesh adaptation: (**a**) Signed distance α and isoline $\alpha = 0$. (**b**) Adapted computational mesh (**c**) Smoothed Heaviside H_ε with $\varepsilon = R/100$.

Instead of computing the distance from a point x to each element defining ω, the idea behind the octree is to select elements located near x, and to compute the distance from them only. The distance computation algorithm is discussed in the following, with use of the nomenclature defined in Table 1.

Table 1. Nomenclature used to discuss distance computation algorithm.

Variable Name	Signification
x	Point of computational mesh
ω	Shape immersed represented by elements
E_x	Closest element of the set representing ω from x
OL_x	Closest octree leaf from x
d_x	Maximal theoretical distance from x to E_x
C_x	Circle/sphere of center x and radius d_x
α_x	Signed distance from x to E_x

All starts with the determination of the octree leaf OL_x which is the closest from x. From the definition of the octree, OL_x is proven not to be empty. Even if the closest element from x, named E_c, is not imperatively stored inside OL_x, its distance to x is inferior or equal to the distance from x to the closest element located inside OL_x. A well-parametrized octree guarantees that the size of the set of elements contained inside a leaf is reasonable. The distances from x to the bounding boxes of every element contained inside OL_x are then computed. The distance to the furthest point of every bounding box is computed, and the minimum obtained is selected. This minimal distance d_x defines a circle/sphere C_x of center x and of radius d_x, in which the closest elements is compulsorily located. The octree is then browsed to determine all the leaves it intersects, which are candidates to host E_x. The bounding boxes of all the elements located in the selected leaves are browsed, and if the minimum distance from x to it is inferior to d_x, the distance from x to the element is computed. α_x is then obtained by selecting the minimum among the distances to elements evaluated.

Octree has been defined to be computationally efficient and stand-alone, and the use of bounding boxes is a key factor to that extent. Large computational savings are enabled as the octree only knows the elements as bounding boxes and, until the very end of the algorithm, distances computed are between x and the boxes. The number of distances from x to the elements, which can be very expensive computationally, is limited to the elements whose bounding box intersect C_x. Browsing all the boxes contained inside OL_x to determine d_x might seem unnecessary, but if this procedure is not completed, the maximal theoretical distance to E_x is the distance to the furthest point of OL_x. Overlooking the boxes enables to reduce the span of C_x, which may translate to a smaller selection of octree leaves and to a reduced number of distances from x to elements. The computational cost of this stage, implying few distance computations to bounding boxes, often tends to be worth the savings brought by the optimization of C_x. The usage of bounding boxes also bring easy generalization of the octree procedure. The selection of the closest elements, to which distance from x is evaluated, is totally independent on the type of elements used. Heterogeneous sets can even be used, with, e.g., facets and fibers mixed.

Figure 7a presents the refined octree defined in Figure 3, where all the leaves of the computational tree are colored in red. To compute the distance from a point P to ω, OL_P is determined and drawn in green in Figure 7b. All the bounding boxes of fibers immersed in this leaf are browsed to determine d_P and C_P. The octree leaves intersecting this circle are determined and asterisked in Figure 7c. The intersection between the bounding boxes of fibers contained in those leaves and C_P is examined, and if, and only if, an intersection is found, the distance from x to the fiber is determined. The same procedure is followed for points Q and R. Those three examples depict the efficiency of the method in different situations (the most frequent situation is the one described by the point R), where the number of evaluations of distances to elements is largely reduced. Table 2 shows a large decrease despite the low number of fibers immersed, which reduces the efficiency of the method. The octree construction and closest leaves determination costs are not included in this situation. However, the recursive construction and the distance to bounding boxes determination are cheap computationally compared to the distance to fibers evaluation, which requires projections. When a deeper octree is used for much bigger ω, evaluating distances to fibers become quite expensive, and savings brought by the octree rise rapidly.

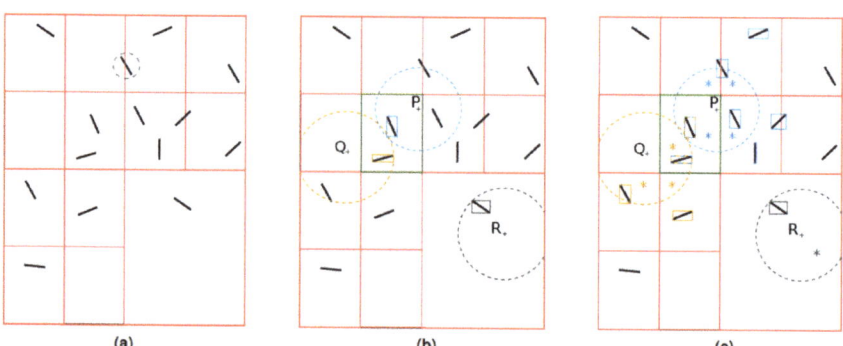

Figure 7. Octree fiber immersion optimization example: (**a**) Final octree. (**b**) Determination of closest leaf. (**c**) Octree leafs to consider.

Table 2. Reduction in the number of evaluations of distances provided by the octree.

Distances to Fiber/Points	P	Q	R	Total
no octree	14	14	14	42
octree	3	3	1	7

3.2. Parallel Anisotropic Mesh Adaptation

Octree-optimized mesh immersion procedure is an efficient way to represent geometries if an accurate computational mesh is used as Section 3.1 stated. The results obtained with this technique are highly dependent on the position of the points, particularly at the interfaces. To that extent, a coupling between mesh immersion and the automatic generation of an anisotropic mesh is proposed in order to reduce the size of the problem to be treated. This iterative process starts with a coarse initial mesh, where geometries are immersed and reconstructed using the methods proposed in Section 3.1. A-posteriori error estimator [13,14] evaluates errors from the level-set results at each computational point, using the smoothed Heaviside function H_ε described in Equation (2). In order to generate an anisotropic mesh, a tensor is defined at each point, enabling to measure the errors along each dimension. In other words, at each computational point, the variation of the function H_ε along each direction is observed.

The adaptation relies on a uniform distribution of the error along the edges of the mesh in all the computational domain. A metric can be built, which allows to deform the mesh in order to attain uniform error: refinement is performed in the areas where the error is too important, while mesh is coarsened where low error is observed. As H_ε is defined from a hyperbolic tangent, major gradients variation are found around the interfaces while the function is almost constant far from the frontiers. Consequently, around the interfaces, low edges are required to attain errors equivalent to the one obtained with large edges where gradients are almost null. Consequently, the new mesh will feature more nodes in the interest zones, and the reconstruction will gain precision. As the metric is built as a tensor, different stretching factors are used for each direction, which guarantees anisotropic meshing.

After several iterations, the errors are uniformly dispersed in the computational domain. Nodes are mostly concentrated around Γ, and the immersed geometry is well described. Highly-stretched mesh cells can be found in regions where very thin description is needed in one dimension while the others do not require particular attention. However, the stretching ratio of the mesh cells is limited, in order to ensure convergence of computations. The automatic and anisotropic mesh adaptation brings versatility, and at the same time guarantees that the results obtained with the mesh immersion procedure will be accurate. The reduction in the number of points required for the reconstruction enables to reduces both memory usage and computational costs. Coupled with an octree, an efficient optimization of the reconstruction is obtained. Moreover, this reconstruction process is executed on a multi-cores context in order to be able to combine the optimizations related to the use of mesh adaptation and octree with massively parallel computing. The parallelization of the process is performed by an iterative coupling between operations of independent adaptive mesh in different partitions and displacement of the interface between these partitions [15,16].

3.3. Weak Scalability Test of the Proposed Reconstruction Approach

To determine the scaling capability of the whole reconstruction procedure, weak scaling tests have been performed on the western French region, Pays de la Loire cluster Liger (a BULL/Atos DLC720 cluster, 6384 cores Intel Xeon (Haswell and Cascade Lake) (compute and visualization parallel procedures), a total of 36,608 Gigabytes of system memory, 5.33 GB per core, FDR Infiniband interconnect (56 GB/s)). Five microstructures were generated, as described previously, while keeping the same geometrical characteristics of fibers. To realize tests with similar workload per processor, the size of the computational domain and the number of immersed fibers were proportionally increased according to the number of the used cores, as detailed in Table 3.

Table 3. Simulation parameters for weak scalability test performed on liger supercomputer.

	Number of Fibers	Domain Edge Size	Number of Cores	Total Mesh Nodes
test 1	8	0.178	1	172,245
test 2	216	0.534	27	728,895
test 3	1000	0.890	125	37,153,365
test 4	4096	1.425	512	160,374,769
test 5	8000	1.781	1000	317,813,266

The reconstruction process started from an initial coarse mesh and took 30 iterations with constant precision and fixed octree parameters. For the different test cases, an average number of mesh nodes per core equal to 3×10^5 was maintained with the exception of test 1 (1.8×10^5 nodes) where the volume of fibers that extend outside the computational domain and are therefore sliced is significant, so leading to a decrease in the number of nodes. Total time of the immersion and adaptation process as a function of the number of cores is represented in Figure 8. For an ideal weak scale test, the run time is expected to stay constant while the workload is increased in direct proportion to the number of processors. For real case, as shown in Figure 8, a deviation can be observed due to communications and partitioning efforts. However, according to the same figure, the running time variation is relatively small between the tests (except for the first one where the workload is different) which allows to consider that for a scaled problem size, the domain reconstruction approach has good efficiency in terms of weak scalability.

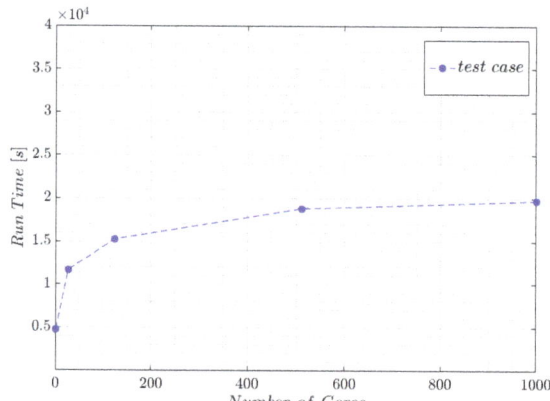

Figure 8. Total reconstruction time evolution as a function of used cores for the different test cases.

4. Flow Simulation Examples: Application to Permeability Computation
4.1. Flow Simulation

The resulting mesh from the reconstruction process can be used to simulate various physical phenomena, such as those involved in fluid-structure interaction problems. Generally, for composite flow applications, incompressible Stokes flow around the fibers is considered. By considering a stationary regime and neglecting the volume forces, the variational form of the Stokes problem for velocity field, **u**, and pressure field, p, is written:

$$(\mathbf{v}, q) \in \mathcal{V}_0 \times \mathcal{Q}$$
$$\begin{cases} (2\eta \varepsilon(\mathbf{u}) : \varepsilon(\mathbf{v}))_\Omega - (p, \nabla \cdot \mathbf{v})_\Omega = 0 \\ (\nabla \cdot \mathbf{u}, q)_\Omega = 0 \end{cases} \quad (4)$$

where ε is the strain rate tensor.

A monolithic approach is used, i.e., the flow Equation (4) are solved on the single mesh defined over the whole computational domain, Ω, regardless of the type of phase it contains. The different phases are distinguished by their physical properties which are taken into account through a mixing law. A linear mixture relation is used for the viscosity, η, and described by the Equation (5).

$$\eta = \eta_f H_\epsilon + \eta_s(1 - H_\epsilon) \tag{5}$$

η_f and η_s are, respectively, the viscosities of the liquid and solid phases. η_s acts as a penalty parameter: when it is high enough, shear rate in the penalized phase becomes close to zero and we find a rigid body motion. This is a simple way to obtain results similar to those provided by an augmented Lagrangian method where a Lagrange multiplier is used to impose a constraint on the solid phase to avoid its deformation [17]. To solve the system (4) using a finite element method, a stabilized approach of VMS type is employed [12]. The used software in this work is ICI-tech, developed at the High Performance Computing Institute (ICI) of Centrale Nantes and implemented for massively parallel context.

4.2. Permeability Computation Procedure

Predicting permeability is a very important issue in the field of composite forming process. However, it is tricky and complex to obtain experimentally and numerically reliable results, because most simulations are carried out in small periodic representative elementary volumes, under a lot of simplifying assumptions that idealize the real media. Here, we chose to rise to the challenge to numerically determine the permeability tensor of a large virtual sample of fibrous media that imitates sophisticated real media. In three-dimensional cases, permeability is characterized by a symmetric second-order tensor **K**. This tensor relates the average fluid velocity $\langle \mathbf{u} \rangle$ to the average pressure gradient on the fluid domain $\langle \nabla p \rangle^f$, as shown by the Darcy law below:

$$\langle \mathbf{u} \rangle = -\frac{\mathbf{K}}{\eta} \langle \nabla p \rangle^f \tag{6}$$

Using a monolithic approach with finite element discretization, the homogenized velocity and pressure fields are written as the sum of their integration on each mesh element Ω_e of the simulation domain Ω:

$$\langle \mathbf{u} \rangle = \frac{1}{V_\Omega} \sum_e \int_{\Omega_e} (1 - H_\epsilon(\alpha)) \mathbf{u} \, d\Omega_e \tag{7}$$

$$\langle \nabla p \rangle^f = \frac{1}{V_{\Omega_f}} \sum_e \int_{\Omega_e} (1 - H_\epsilon(\alpha)) \nabla p \, d\Omega_e \tag{8}$$

where V_Ω is the volume of the total domain and V_{Ω_f} is the volume of the fluid domain.

To predict permeability, the proposed simulation procedure relies on microstructure generation, phase reconstruction, mesh adaptation, and resolution of the Stokes equations, considering that fibers are static and impermeable. In fact, to determine all components of **K**, three flows in the three directions x, y, and z are successively simulated, an exponent $\{1, 2, 3\}$ is referred to each one. The flow is induced by an imposed pressure gradient. Depending on the direction where the flow is desired, a constant pressure field on the input face of the simulation domain against a null field on the output face is imposed. For the other faces of the domain, only the normal component of the velocity field is imposed as null. Assuming that the permeability tensor is symmetric and positive definite, its components can be calculated by the resolution of the overdetermined linear system given by:

$$\begin{bmatrix} \langle\nabla p_x\rangle^1 & \langle\nabla p_y\rangle^1 & \langle\nabla p_z\rangle^1 & 0 & 0 & 0 & 0 & 0 & 0 \\ 0 & 0 & 0 & \langle\nabla p_x\rangle^1 & \langle\nabla p_y\rangle^1 & \langle\nabla p_z\rangle^1 & 0 & 0 & 0 \\ 0 & 0 & 0 & 0 & 0 & 0 & \langle\nabla p_x\rangle^1 & \langle\nabla p_y\rangle^1 & \langle\nabla p_z\rangle^1 \\ \langle\nabla p_x\rangle^2 & \langle\nabla p_y\rangle^2 & \langle\nabla p_z\rangle^2 & 0 & 0 & 0 & 0 & 0 & 0 \\ 0 & 0 & 0 & \langle\nabla p_x\rangle^2 & \langle\nabla p_y\rangle^2 & \langle\nabla p_z\rangle^2 & 0 & 0 & 0 \\ 0 & 0 & 0 & 0 & 0 & 0 & \langle\nabla p_x\rangle^2 & \langle\nabla p_y\rangle^2 & \langle\nabla p_z\rangle^2 \\ \langle\nabla p_x\rangle^3 & \langle\nabla p_y\rangle^3 & \langle\nabla p_z\rangle^3 & 0 & 0 & 0 & 0 & 0 & 0 \\ 0 & 0 & 0 & \langle\nabla p_x\rangle^2 & \langle\nabla p_y\rangle^2 & \langle\nabla p_z\rangle^3 & 0 & 0 & 0 \\ 0 & 0 & 0 & 0 & 0 & 0 & \langle\nabla p_x\rangle^3 & \langle\nabla p_y\rangle^3 & \langle\nabla p_z\rangle^3 \\ 0 & 1 & 0 & -1 & 0 & 0 & 0 & 0 & 0 \\ 0 & 0 & 1 & 0 & 0 & 0 & -1 & 0 & 0 \\ 0 & 0 & 0 & 0 & 0 & 1 & 0 & -1 & 0 \end{bmatrix} \begin{bmatrix} K_{xx} \\ K_{xy} \\ K_{xz} \\ K_{yx} \\ K_{yy} \\ K_{yz} \\ K_{zx} \\ K_{zy} \\ K_{zz} \end{bmatrix} = -\eta \begin{bmatrix} \langle u_x\rangle^1 \\ \langle u_y\rangle^1 \\ \langle u_z\rangle^1 \\ \langle u_x\rangle^2 \\ \langle u_y\rangle^2 \\ \langle u_z\rangle^2 \\ \langle u_x\rangle^3 \\ \langle u_y\rangle^3 \\ \langle u_z\rangle^3 \\ 0 \\ 0 \\ 0 \end{bmatrix} \quad (9)$$

The solution obtained from the resolution of this matrix system (9) is, obviously, an approximate solution. To ensure a perfect symmetry of **K**, if necessary, the following modification to the extra diagonal terms is made:

$$K_{ij}^{final} = K_{ji}^{final} = \frac{K_{ij} + K_{ji}}{2} \quad (10)$$

4.3. Permeability Computation Validation

To validate permeability computation, the whole procedure was applied to a parallel square packing of fibers having an identical diameter. Rigidity of fibers was ensured by imposing $\eta_s = 500\eta_f$ and a zero velocity condition was imposed upon them. Figure 9a shows the used geometry configuration for $V_f = 25.65\%$. Equation (11) represents its calculated permeability tensor adimensionalized by the square of fiber radius which respect a transverse isotropic form as expected from the symmetry of the packing.

$$\mathbf{K} = \begin{bmatrix} 0.21 & 0 & 0 \\ 0 & 0.21 & 0 \\ 0 & 0 & 0.28 \end{bmatrix} \quad (11)$$

Permeability evolution according to fiber volume fraction was studied by varying fiber diameter and keeping same the domain size for all simulations. The obtained results of normalized transverse permeability are reported in Figure 9b and compared to the model of [18–20]. The observed permeability values through this graph are in the same order than the one obtained from analytical laws which is relevant to our approach.

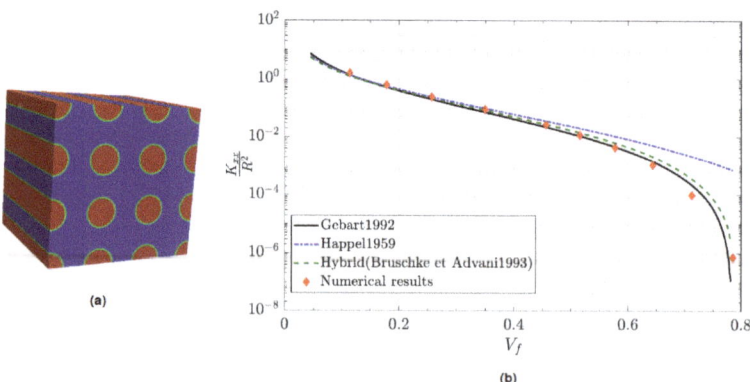

Figure 9. Comparison of computed permeability with some analytical models: (**a**) Simulated parallel square packing configuration. (**b**) Normalized transverse permeability results.

4.4. Application for 10,000 Fibers

4.4.1. Microstructure Generation

The first step of the process is the microstructure generation using the octree optimized algorithm described in Section 2. A sample of approximately 10,000 (exactly 10,062) collision-free fibers is created in a cubic domain with an edge length of 1.35 mm. The fibers have a common diameter of 15 µm and a length that follows a normal distribution of mean 0.2 mm and standard deviation 0.03 mm. The obtained volume fraction is $V_f = 14\%$. The orientation state is nearly isotropic and is given by the following orientation tensor a_2 [21]:

$$a_2 = \begin{bmatrix} 0.334241 & -0.00219696 & -0.018116 \\ -0.00219696 & 0.34166 & -0.00620966 \\ -0.018116 & -0.00620966 & 0.324099 \end{bmatrix}$$

Figure 10 shows the set of the generated fibers. Despite the fact that the generation is sequential, these fibers are created in only **1min44s** thanks to the octree contribution.

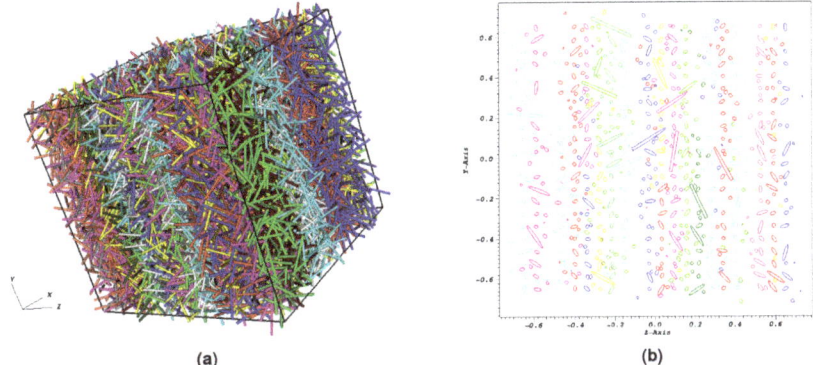

(a)   (b)

Figure 10. Studied generated microstructure: (**a**) 10,000 generated fibers. (**b**) A random slice showing no collisions.

4.4.2. Microstructure Reconstruction with Adaptative Mesh

The computation was performed on 384 cores. Starting from an initial mesh of ≈4.6 million nodes and ≈27 million elements, after 30 iterations, an adapted final mesh of ≈67 million nodes and ≈391 million elements is created by the methods described in Sections 3.1 and 3.2. For H_ε with ε = 3.125 µm, the total immersion and adaptation process required **4h52min** for the 30 iterations. Figure 11 shows the evolution, in a number of elements for each iteration of the mesh adaptation, as well as the computational time. During the first iterations of immersion of the generated fibers in the initial mesh, the mesher adds a considerable number of elements until reaching a peak at the ninth iteration, in order to properly capture the geometries of all the fibers at first. Then, the mesher focuses its work on optimizing the mesh adaptation at the interfaces while respecting a criterion of mesh quality. Once an efficient mesh is achieved, the number of elements stabilizes. The time evolution curve naturally follows the evolution of the mesh size.

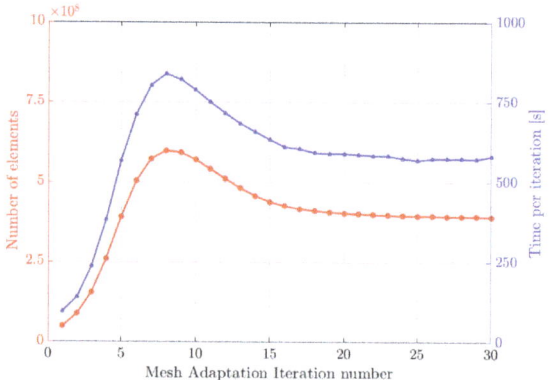

Figure 11. Evolution of the mesh number of elements (**left** axis) and calculation cost (**right** axis) during the 30 iterations of adaptation of anisotropic mesh, performed on 384 cores

4.4.3. Flow Resolution and Permeability Tensor Computation

Three pressure gradients are applied to the constructed finite element mesh in order to generate the flows required for the identification of **K**. Figure 12 shows the pressure field and velocity vectors around the immersed fibers for the flow in the x direction. These results were obtained for a resolution time of the system (4) equal to approximately **7min** minutes on 384 CPUs.

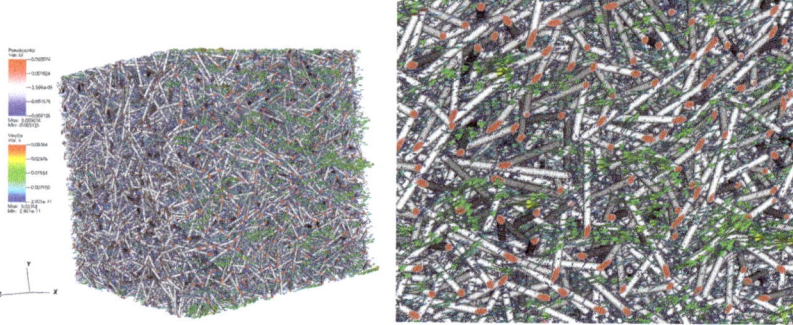

Figure 12. Flow according to x direction: (**a**) velocity vector around the fibers. (**b**) Zoom around a zone of the figure.

The predicted full permeability tensor adimensionalized by the square of fiber radius for this media is as follows:

$$\mathbf{K} = \begin{bmatrix} 0.7322 & -0.0013 & -0.0033 \\ -0.0013 & 0.7444 & -0.0025 \\ -0.0033 & -0.0025 & 0.7089 \end{bmatrix}$$

For isotropic material, only the three diagonal elements are non-null and they are equal. Here, the studied sample is nearly isotropic. For this reason, the obtained diagonal elements are quite similar and the off-diagonal elements are smaller by around two orders of magnitude.

5. Conclusions

Obtained results show our capability thanks to an octree implementation to deal with big data in terms of input of permeability simulation and to perform reliable finite element calculation on complex geometries. Through the proposed method, further studies can be conducted to better quantify the impact of the microstructural parameters on the permeability and, thus, avoiding problems related to the choice of the size of the simulation domains, which remains rather delicate to define, especially in the case of non-periodic geometries. We can also think about exploring the permeability of multiaxial tissues of the non-crimp fabric (NCF) or textile type. Thanks to the several numerical optimization, the permeability can thus be evaluated at the microscopic scale on several layers by representing the fibers inside the wicks.

Author Contributions: Conceptualization and methodology, N.A., L.D., L.S., H.D. and E.A.-C.; software, N.A., L.D., H.D. and P.L.; validation, N.A., L.S., H.D. and E.A.-C.; writing—original draft preparation, N.A. and L.D.; writing—review and editing, N.A., L.S. and E.A.-C. All authors have read and agreed to the published version of the manuscript.

Funding: This research received no external funding.

Acknowledgments: This work was performed by using HPC resources of the Centrale Nantes Supercomputing Centre on the cluster Liger. Additionally, we would like to warmly thank GENCI and TGCC French National Supercomputing Facility for CPU time under projects A00*0607575 allowing us to test the approach presented in this article on Joliot-Curie.

Conflicts of Interest: The authors declare no conflict of interest.

References

1. Samet, H. *The Design and Analysis of Spatial Data structuresAddison-Wesley Series in Computer Science*; Addison-Wesley: Reading, MA, USA, 1990.
2. Szeliski, R. Rapid octree construction from image sequences. *CVGIP Image Underst.* **1993**, *58*, 23–32. [CrossRef]
3. Laurmaa, V.; Picasso, M.; Steiner, G. An octree-based adaptive semi-Lagrangian VOF approach for simulating the displacement of free surfaces. *Comput. Fluids* **2016**, *131*, 190–204. [CrossRef]
4. Fan, W.; Wang, B.; Paul, J.C.; Sun, J. An octree-based proxy for collision detection in large-scale particle systems. *Sci. China Inf. Sci.* **2013**, *56*, 1–10. [CrossRef]
5. Kim, Y.H.; Ko, S.L. Improvement of cutting simulation using the octree method. *Int. J. Adv. Manuf. Technol.* **2006**, *28*, 1152–1160. [CrossRef]
6. Wang, H.Y.; Liu, S.G. A collision detection algorithm using AABB and octree space division. In *Advanced Materials Research*; Trans Tech Publications: Zurich, Switzerland, 2014; Volume 989, pp. 2389–2392.
7. Baehmann, P.L.; Wittchen, S.L.; Shephard, M.S.; Grice, K.R.; Yerry, M.A. Robust, geometrically based, automatic two-dimensional mesh generation. *Int. J. Numer. Methods Eng.* **1987**, *24*, 1043–1078. [CrossRef]
8. Péron, S.; Benoit, C. Automatic off-body overset adaptive Cartesian mesh method based on an octree approach. *J. Comput. Phys.* **2013**, *232*, 153–173. [CrossRef]
9. Zhang, Y.; Bajaj, C.; Sohn, B.S. 3D finite element meshing from imaging data. *Comput. Methods Appl. Mech. Eng.* **2005**, *194*, 5083–5106. [CrossRef] [PubMed]
10. Mezher, R. Modeling and Simulation of Concentrated Suspensions of Short, Rigid and Flexible Fibers. Ph.D. Thesis, Engineering Sciences [physics], Ecole Centrale de Nantes (ECN), Nantes, France, 2015.
11. Aissa, N. Simulation Haute Performance des Suspensions de Fibres Courtes Pour les Procédés de Fabrication de Composites. Ph.D. Thesis, Mécanique des Fluides [physics.class-ph], École Centrale de Nantes, Nantes, France, 2021.
12. Coupez, T.; Hachem, E. Solution of high-Reynolds incompressible flow with stabilized finite element and adaptive anisotropic meshing. *Comput. Methods Appl. Mech. Eng.* **2013**, *267*, 65–85. [CrossRef]
13. Coupez, T. Metric construction by length distribution tensor and edge based error for anisotropic adaptive meshing. *J. Comput. Phys.* **2011**, *230*, 2391–2405. [CrossRef]
14. Zhao, J.X.; Coupez, T.; Decencière, E.; Jeulin, D.; Cárdenas-Peña, D.; Silva, L. Direct multiphase mesh generation from 3D images using anisotropic mesh adaptation and a redistancing equation. *Comput. Methods Appl. Mech. Eng.* **2016**, *309*, 288–306. [CrossRef]
15. Digonnet, H.; Coupez, T.; Laure, P.; Silva, L. Massively parallel anisotropic mesh adaptation. *Int. J. High Perform. Comput. Appl.* **2019**, *33*, 3–24. [CrossRef]
16. Silva, L.; Coupez, T.; Digonnet, H. Massively parallel mesh adaptation and linear system solution for multiphase flows. *Int. J. Comput. Fluid Dyn.* **2016**, *30*, 431–436. [CrossRef]
17. Beaume, G. Modélisation et Simulation Numérique de l'Écoulement d'un Fluide Complexe. Ph.D. Thesis, Mécanique [physics.med-ph], École Nationale Supérieure des Mines de Paris, Paris, France, 2008.
18. Gebart, B. Permeability of unidirectional reinforcements for RTM. *J. Compos. Mater.* **1992**, *26*, 1100–1133. [CrossRef]
19. Happel, J. Viscous flow relative to arrays of cylinders. *AIChE J.* **1959**, *5*, 174–177. [CrossRef]

20. Bruschke, M.; Advani, S. Flow of generalized Newtonian fluids across a periodic array of cylinders. *J. Rheol.* **1993**, *37*, 479–498. [CrossRef]
21. Advani, S.G.; Tucker, C.L., III. The use of tensors to describe and predict fiber orientation in short fiber composites. *J. Rheol.* **1987**, *31*, 751–784. [CrossRef]

Article

A Data-Driven Space-Time-Parameter Reduced-Order Model with Manifold Learning for Coupled Problems: Application to Deformable Capsules Flowing in Microchannels

Toufik Boubehziz [1], Carlos Quesada-Granja [1], Claire Dupont [1], Pierre Villon [2], Florian De Vuyst [3,*] and Anne-Virginie Salsac [1]

[1] Biomechanics and Bioengineering Laboratory (UMR CNRS 7338), Université de Technologie de Compiègne CNRS, Alliance Sorbonne Université, 60203 Compiègne, France; toufik.boubehziz@utc.fr (T.B.); quesadagranja@gmail.com (C.Q.-G.); claire.dupont@utc.fr (C.D.); a.salsac@utc.fr (A.-V.S.)

[2] Laboratoire Roberval, Université de Technologie de Compiègne, Alliance Sorbonne Université, 60203 Compiègne, France; pierre.villon@utc.fr

[3] Laboratoire de Mathématiques Appliquées de Compiègne (EA 2222), Université de Technologie de Compiègne, Alliance Sorbonne Université, 60203 Compiègne, France

* Correspondence: florian.de-vuyst@utc.fr

Abstract: An innovative data-driven model-order reduction technique is proposed to model dilute micrometric or nanometric suspensions of microcapsules, i.e., microdrops protected in a thin hyperelastic membrane, which are used in Healthcare as innovative drug vehicles. We consider a microcapsule flowing in a similar-size microfluidic channel and vary systematically the governing parameter, namely the capillary number, ratio of the viscous to elastic forces, and the confinement ratio, ratio of the capsule to tube size. The resulting space-time-parameter problem is solved using two global POD reduced bases, determined in the offline stage for the space and parameter variables, respectively. A suitable low-order spatial reduced basis is then computed in the online stage for any new parameter instance. The time evolution of the capsule dynamics is achieved by identifying the nonlinear low-order manifold of the reduced variables; for that, a point cloud of reduced data is computed and a diffuse approximation method is used. Numerical comparisons between the full-order fluid-structure interaction model and the reduced-order one confirm both accuracy and stability of the reduction technique over the whole admissible parameter domain. We believe that such an approach can be applied to a broad range of coupled problems especially involving quasistatic models of structural mechanics.

Keywords: data-driven model; model order reduction; proper orthogonal decomposition; manifold learning; diffuse approximation; microcapsule suspension; Hausdorff distance

1. Introduction

Numerical modeling and simulation today appear to be an indispensable science to analyze physics-coupled problems (e.g., micrometric and nanometric suspensions), but also for innovative design and optimization of complex three-dimensional systems in engineering and industry (health, automotive, aircraft, etc.). Although one can nowadays find robust and accurate open-source or commercial codes for the simulation of multiphysics systems, it is still hard to use them in the context of robust or optimal design because of the prohibitive computational time that does not match with engineering production horizons. In order to accelerate computations, one can make use of parallel HPC (High Performance Computing) facilities, but this can become financially costly. For most applications, even with HPC facilities, the evaluation of the solutions takes days or weeks for three-dimensional multi-coupled problems.

Alternatively, a current tendency is to use machine learning or artificial intelligence tools to capitalize knowledge stored into data and use it for future case studies. It leads to less redundant or useless computations, while the database of results continues to grow with more and more relevant information contents. With machine learning, one can expect to explore the design spaces in an easier, faster and more efficient way. However, one usually needs expertise to design and train artificial neural networks (ANN) correctly. For high-dimensional data, the training stage may require large computational resources and issues of quality of data may also be raised. Machine learning can be extended to time dependent problems and dynamical system using e.g., recurrent neural networks [1].

However, for particular use cases like physical systems, machine learning algorithms are designed to return three-dimensional spatial fields. This means that the outputs of the networks are high-dimensional vectors, which may induce training convergence and accuracy issues. Over the last two years, one could observe the rise of so-called Physics-informed neural networks (PINN) where the artificial neural networks are trained from a loss function that includes physical information (like partial differential equations), see e.g., [2].

Another class of 'machine learning' methods are the data-driven model order reduction (MOR) techniques, that use data generated from a (time-consuming) high-fidelity solver, also called the full-order model (FOM) [3,4]. Data-driven and non-intrusive reduced-order models (ROM) can be seen as supervised ANN [5,6]. For parametrized partial differential problems, ROMs usually perform a dimensionality reduction by means of suitable reduced bases. This can be achieved via different approaches such as the Proper Orthogonal Decomposition (POD) [7–9], piecewise tangential interpolation [10], Proper Generalized Decompositions (PGD) [11–13], Empirical Interpolation Methods (EIM) [14–16] or via different greedy procedures [17]. Then one has to find the manifold that maps the parameters to the coefficients of linear combination of the reduced basis functions. This can also be achieved in a supervised way, by means of universal approximation techniques like diffuse approximation [18] for example. Using ROMs may lead to substantial speedups as compared to FOM, from say 10 up to 10,000. One can even imagine real-time computations in some cases [19].

The 'ultimate' case is that of space-time-parameter problems involving spatial fields, timeline and design variables. This is of course of industrial importance, but still an issue and a current active field of research (see [20] for example). For such problems, the data are generally organized in data cubes (Figure 1). In this paper, a data-driven reduced-order modeling approach is proposed for space-time-parameter mechanical problems involving an equation of kinematics and a quasi-static law of equilibrium. As particular application, the physical problem that is addressed is the dynamics of dilute suspensions of micrometric capsules in microfluidic channels. Microcapsules can be used in Healthcare as innovative drug transportation vehicles into blood vessels and are expected to deliver drugs at identified targets [21,22]. They are composed of an elastic membrane protecting a liquid inner core and are used in suspension in another liquid. Testing them in microfluidic environments offers great potential to determine the capsule behavior and characterize the mechanical properties of the membrane [23–29], but also for sorting or enrichment of capsule suspensions [30–34]. We presently focus on the flow of a dilute suspension of initially spherical micrometric capsules in a microfluidic channel, which is a complex three-dimensional inertialess fluid-structure interaction problem that interestingly depends on only two independent design variables: the capillary number of the capsule Ca, which is a non-dimensional number that estimates the order of magnitude of the viscous forces acting on the capsule with respect to the elastic forces that build up in the membrane, and the confinement ratio a/ℓ that provides a comparison between the initial capsule diameter and the channel width.

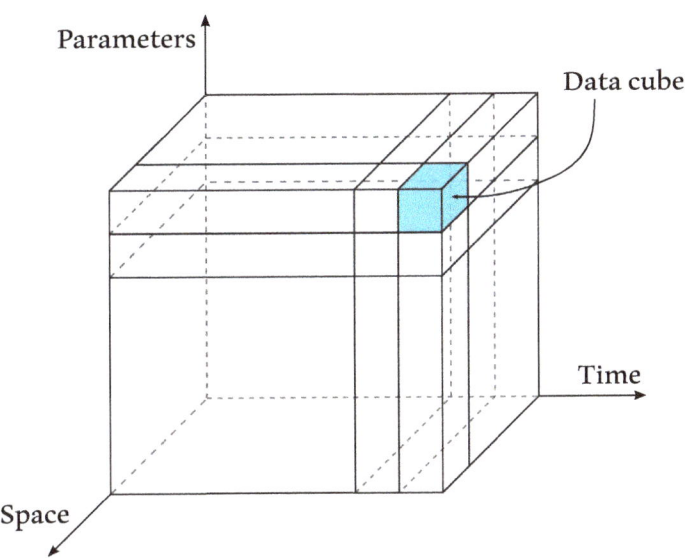

Figure 1. Space-time-parameter data cube.

Proper Orthogonal Decomposition has been shown to be particularly suitable to build reduced order models (ROM) of microcapsules [35], but so far no model capable of predicting capsule dynamics currently exists. The originality of the paper is to propose a ROM of the capsule-fluids interactions which provides the time-evolution of the capsule shape for any parameter values. From the capsule shape, it is indeed possible to deduce all the quantities of interest (viscous load, internal tensions within the membrane, membrane energy, etc.) in post-treatment. The ROM is inspired from the physical problem, in which the boundary condition stipulates that the fluid velocity equals the capsule membrane velocity. The correction of the capsule node position field can thus be obtained by integrating the velocity field over time. The challenge remains to correlate the position and velocity fields, which we propose to do with diffuse approximation and manifold learning [18,36,37] using the principal modes of both fields obtained by POD decomposition.

Numerical experiments will demonstrate the accuracy and efficiency of the approach, comparing reduced-order solutions to the full-order ones. We believe that the methodology proposed in this paper can be applied to a broad range of multiphysics problems such as fluid-structure interactions, structural dynamics using quasi-static structural mechanics models and related problems.

This paper is organized as follows. In Section 2, we describe the physical problem and its full order model solution. In Section 3.2, we construct parametric and spatial reduced-order modes using a set of pre-computed simulations. This allows to introduce a reduced-order model that expresses the displacement and velocity of a capsule at a selection of snapshots. In Section 3.4, we build a reduced model that corresponds to any parameter vector of Ca and a/ℓ values by estimating its corresponding principal components. Then, with the use of a Diffuse Approximation (DA) method, we adopt a data-driven manifold learning to predict the deformation of the capsule in the flow for a chosen time discretization. Finally, in Section 4, we will validate the whole computational ROM approach with a comparison to the FOM solutions.

2. Material and Methods

2.1. Problem Statement

Let us consider a spherical micrometric capsule of radius a freely placed in a three-dimensional microfluidic channel with square cross-section of length 2ℓ (see Figure 2). The capsule and the channel are filled with an incompressible Newtonian fluid of the same constant density ρ and dynamic viscosity μ. The capsule is enclosed by a thin hyperelastic isotropic membrane (surface shear modulus G_s, area expansion modulus $K_s = 3G_s$). It is subjected to a Poiseuille flow of mean velocity V.

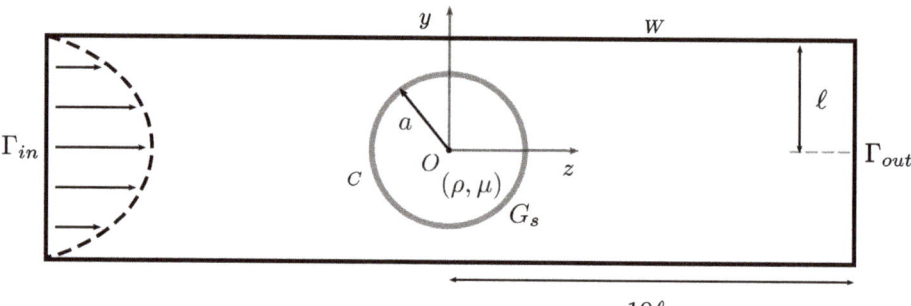

Figure 2. Initial configuration considered in the FOM model: an initially spherical capsule is placed at the center of a square-section channel. The time-evolution of its dynamics is computed using a reference frame centred onto the capsule centre of mass.

The problem is governed by two dimensionless numbers:

- The confinement ratio a/ℓ, ratio of the capsule to tube sizes;
- The capillary number $Ca = \mu V/G_s$, ratio of the viscous forces onto the capsule membrane to the membrane elastic forces.

The Reynolds number of the external flow is assumed to be very small (typically of order 10^{-2} or less), inertia being negligible because of the spatial scales involved in the problem. As far as the internal flow is concerned, its velocity is induced by the motion of the capsule membrane, which is itself entrained by the external flow: it is thus of smaller amplitude than that of the external flow. Hence, the flow in the internal ($\beta = in$) and external ($\beta = ex$) fluids are described by the Stokes equations:

$$\nabla \cdot v^\beta = 0, \ \nabla \cdot \sigma^\beta = 0, \ \beta = in, ex. \tag{1}$$

where σ^β is stress tensor in the fluids. The Stokes equations are defined in the domains bounded by the capsule membrane for $\beta = in$ and by the capsule membrane and the channel wall for $\beta = ex$. The inlet Γ_{in} and outlet Γ_{out} cross-sections of the channel are assumed to be far from the capsule (10ℓ in the FOM model). The reference frame (O, x, y, z) is fixed on the capsule center of mass O at each time step. For the velocity vector field v^β and the pressure field p^β, we consider the following boundary conditions (note that the boundary conditions include wall confinement effects, see [38] for more details):

- The flow perturbation induced by the capsule vanishes at Γ_{in} and Γ_{out}:

$$v^{ex}(x,t) \to v^\infty(x), \text{ when } x \in \Gamma_{in} \cup \Gamma_{out}, \tag{2}$$

where v^∞ is the Poiseuille flow velocity of the suspending fluid in the absence of capsule. For a square channel we have the expression in expansion form

$$v^\infty(x,y) = \frac{\sum_{n=1,3,\ldots}^{\infty} \frac{\pi V}{n^3} \left[1 - \frac{\cosh(n\pi x/\ell)}{\cosh(n\pi/2)}\right] \sin(n\pi(y/\ell + 1/2))}{2\left[\frac{\pi^4}{96} - \sum_{n=1,3,\ldots}^{\infty} \frac{\tanh(n\pi/2)}{n^5 \pi/2}\right]}. \tag{3}$$

- Uniform pressure at Γ_{in} and Γ_{out}:

$$p^{ex}(x,t) = 0, \ x \in \Gamma_{in}, \tag{4}$$

$$p^{ex}(x,t) = \Delta p(t) + \Delta p^\infty, \ x \in \Gamma_{out}, \tag{5}$$

where Δp^∞ is the undisturbed suspending pressure drop in the absence of capsule and Δp is the additional pressure drop due to the capsule.
- No slip boundary conditions on the channel wall W:

$$v^{ex}(x,t) = 0, \ \text{for } x \in W \tag{6}$$

- No slip boundary conditions on the capsule membrane C:

$$v^{in}(x,t) = v^{ex}(x,t) = \frac{\partial}{\partial t}x(X,t), \ \text{for } x \in C \tag{7}$$

where $\frac{\partial}{\partial t}x(X,t)$ is the membrane velocity at position x at time t, and X is the reference position vector of the capsule membrane.
- The normal loading continuity indicates that the load q on the membrane is due to the viscous traction jump

$$\left[\sigma^{ex}(x) - \sigma^{in}(x)\right] \cdot n = q, \ \text{for } x \in C \tag{8}$$

where n is the outward unit normal vector.

As the membrane thickness is negligibly small compared to the capsule dimensions, the membrane can be considered as a hyperelastic surface devoid of bending stiffness. The in-plane deformation is then measured by the principal extension ratios λ_1 and λ_2, that measure the in-plane deformation. Owing to the combined effects of hydrodynamic forces, boundary confinement, and membrane deformability, the capsule can be highly deformed. Consequently, the choice of membrane constitutive law is important. We consider the Neo-Hookean (NH) constitutive law that models the membrane as an infinitely thin sheet of a three-dimensional isotropic and incompressible material. It was indeed shown to adequately model microcapsules with a cross-linked proteic membrane [23,24,39]. The principal Cauchy in-plane tensions τ_i ($i = 1, 2$) (forces per unit arc length of deformed surface curves) can be expressed as a function of the principal extension ratios:

$$\tau_1 = \frac{G_s}{\lambda_1\lambda_2}\left[\lambda_1^2 - \frac{1}{(\lambda_1\lambda_2)^2}\right], \ \tau_2 = \frac{G_s}{\lambda_1\lambda_2}\left[\lambda_2^2 - \frac{1}{(\lambda_1\lambda_2)^2}\right]. \tag{9}$$

2.2. Discrete Full Order Model (FOM)

The Fluid-Structure Interaction (FSI) problem is numerically modeled by coupling the Boundary Integral Method (BIM) that solves the fluid Equations (2)–(8) with the Finite Element Method (FEM) that solves the membrane mechanical problem [38,40] using the Caps3D in-house code. The unknowns are the discrete displacement field $\{u\}$ and the discrete velocity field $\{v\}$ at the nodes of the membrane mesh. The equation of kinematics states that $\frac{d}{dt}\{u\} = \{v\}$. The forces exerted onto the membrane are computed by the FEM. The deformation of the membrane is computed from the velocity vector field obtained at the membrane nodes by solving the Stokes equations with the BIM, leading to a nonlinear relation written in abstract form $\{v\} = \{\mathcal{N}\}(\{u\})$. For a given parameter vector $\boldsymbol{\theta} = (\theta_1, \theta_2)^T$, where $\theta_1 = Ca$ and $\theta_2 = a/\ell$, the time-continuous semi-discrete FSI scheme reads in abstract form

$$\frac{d}{dt}\{u\}(t) = \{v\}(t),$$
$$\{v\}(t) = \{\mathcal{N}\}(\{u\}(t), \boldsymbol{\theta}), \quad t \in (0, T_f],$$
$$\{u\}(0) = \{0\}, \; \{v\}(t) = \{\mathcal{N}\}(\{0\}, \boldsymbol{\theta})$$

where $\{u\}(t)$ and $\{v\}(t)$ represent the discrete FE displacement and velocity fields at continuous time t, and T_f is the final time. For time discretization, either a forward Euler scheme or a second order Runge-Kutta scheme is used with a suitable constant time step $\delta t > 0$. The Euler scheme reads

$$\{u^{i+1}\} = \{u^i\} + \delta t \{v^i\},$$
$$\{v^{i+1}\} = \{\mathcal{N}\}(\{u^{i+1}\}, \boldsymbol{\theta}),$$
$$\{u^0\} = \{0\}, \; \{v^0\} = \{\mathcal{N}\}(\{0\}, \boldsymbol{\theta})$$

where $\{u^i\}$ and $\{v^i\}$ represent the discrete FE displacement and velocity fields at discrete time $t^{i,\delta} = i\,\delta t \leq T_f$. For second-order accuracy in time, a Runge-Kutta Ralston scheme is used:

$$\{\hat{u}^{i+2/3}\} = \{u^i\} + \frac{2}{3}\delta t \{v^i\},$$
$$\{\hat{v}^{i+2/3}\} = \{\mathcal{N}\}(\{\hat{u}^{i+2/3}\}, \boldsymbol{\theta}),$$
$$\{u^{i+1}\} = \{u^i\} + \frac{\delta t}{4}\left(\{v^i\} + 3\{\hat{v}^{i+2/3}\}\right),$$
$$\{v^{i+1}\} = \{\mathcal{N}\}(\{u^{i+1}\}, \boldsymbol{\theta}),$$
$$\{u^0\} = \{0\}, \; \{v^0\} = \{\mathcal{N}\}(\{0\}, \boldsymbol{\theta}).$$

Because of the explicit nature of the numerical schemes for the equation of kinematics, the time step is subject to a Courant-Friedrichs-Lewy (CFL)-like stability condition

$$\dot{\gamma}\,\delta t < C\,\frac{\Delta h_C}{\ell}\,Ca, \tag{10}$$

where $\dot{\gamma} = V/\ell$, $C > 0$ is a constant and Δh_C is the typical mesh size (see [40]). In practice, we first use small time steps, and tune them to be big enough but not too close to the stability boundary. This process does not take too much time.

2.3. Design of Experiment, Database of FOM Results

Simulations of the FOM problem have been run varying the governing parameters in the range $[0; 0.2]$ for the capillary number Ca and $[0.75; 1.2]$ for the confinement ratio a/ℓ. For $a/\ell \geq 0.95$, the capsule is initially pre-deformed into an ellipsoid of semi-minor axis equal to 0.9. This pre-deformation does not have any impact on the steady-state capsule shape and is enough to avoid contacts between the capsule membrane and the channel wall. The resulting numerical database, composed of $N_c = 118$ $(Ca, a/l)$ samples (Figure 3), contains the time-evolution of the three-dimensional position (or displacements) vectors of the capsule membrane nodes. Only the configurations for which a steady-state shape has been reached are retained. No steady state is found above the dotted red line of Figure 3, the microcapsules exhibiting continuous elongation owing to the strain-softening behavior of the membrane law [41].

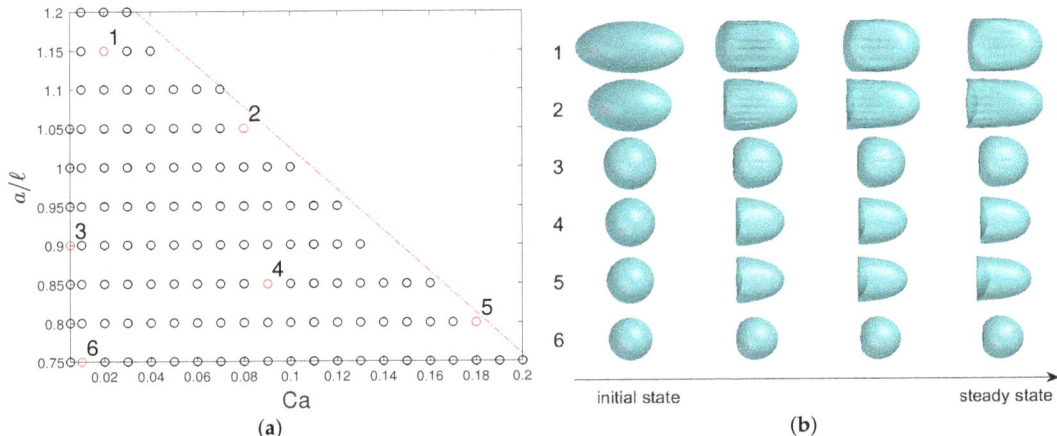

Figure 3. (a) Values of Ca and a/l included in the FOM database in the case of an initially spherical capsule with a Neo-Hookean membrane flowing in a square-section microfluidic channel. The parameter domain where a steady capsule deformation can be reached is delimited by the red dotted line. (b) Time evolution of capsule deformation along the microfluidic channel shown as illustration for the 6 cases indicated with numbers in figure (a). The capsule is pre-deformed into an ellipsoid when $a/\ell \geq 0.95$.

In the ROM model, we consider the capsule positions in the laboratory reference frame (and not the reference frame centred on the capsule centre of mass as in the FOM model). The capsule thus moves along the microchannel. For data generation, we pick up time snapshot solutions at coarser discrete times $t^i = i \Delta t$, where $\Delta t = m \, \delta t$ for some integer $m \geq 1$. The total number of coarse discrete times is denoted N_t. Let $\{X\}_{[n]}$ and $\{x\}_{[n]}(t^i) \in \mathbb{R}^3$, $n = 1, \ldots, N_x$, be the coordinates of node number n of the capsule mesh in the reference configuration (at time $t = 0$) and at discrete time t^i, $i = 1, \ldots, N_t$. The coordinates in the current configuration $\{x\}(t^i)$ are function of the $(Ca, a/\ell)$ parameter value denoted θ_j, $j = 1, \ldots, N_c$. The database is thus stored as a datacube of 3D-space, 1D-time and 2D-parameter data. The displacement vector is then $\{u\}\left(\{X\}_{[n]}, t^i, \theta_j\right) = \{x\}_{[n]}(t^i) - \{X\}_{[n]}$. The velocity vector $\{v\}$ is calculated by finite differences from the position vector. Typically, for a standard capsule FOM simulation, N_x is of order 10^3 and N_c of order 10^2. The time step Δt is chosen such that N_t is of order 10^2.

3. Reduced Order Model (ROM)

Reduced order modeling aims at deriving a lightweight model of low-order dimension from solutions obtained by the FOM, while trying to keep the same order of accuracy. There are many reasons for doing that. In particular, parameter exploration and sensitivity analysis are made easier because of large speedups using the ROM compared to the prohibitive FOM computational time. One can also imagine real-time parameter exploration and visualization of capsule evolution.

3.1. Overview

We first give a general overview of the proposed data-driven model order reduction methodology. The approach is classically made of an offline stage for the search of the principal components and POD coefficient matrices of the FOM solutions, followed by an online stage where a parameter is chosen and a low-order dynamical system is run to get the solutions.

1. **Offline stage.** We reduce the data dimensionality by means of a double POD basis for space and parameter variables. The displacement field is represented as

$$\{u\}(\{X\}, t, \boldsymbol{\theta}) = \sum_{k=1}^{K_u^x} \sum_{\ell=1}^{K_u^c} A_{k\ell}(t) \{\Phi_u^r\}_k (\psi_u(\boldsymbol{\theta}))_\ell, \qquad (11)$$

where $\{\Phi_u^r\}_k \in \mathbb{R}^{3N_x}$ are the spatial POD modes, $\psi_u(\boldsymbol{\theta}) \in \mathbb{R}^{K_u^c}$ the parameter modes and $A_{k\ell}(t)$ scalar coefficients depending on time t. The truncation ranks are K_u^x and K_u^c, respectively (the 'x' superscript stands for 'space' and the 'c' superscript for 'configuration'). We use a similar representation for the velocity field:

$$\{v\}(\{X\}, t, \boldsymbol{\theta}) = \sum_{k=1}^{K_v^x} \sum_{\ell=1}^{K_v^c} B_{k\ell}(t) \{\Phi_v^r\}_k (\psi_v(\boldsymbol{\theta}))_\ell. \qquad (12)$$

The determination of the double POD basis is achieved by singular value decomposition (SVD) from the datacube with different rearrangements of the data in stacked matrix form. The truncation ranks $K_u^x, K_u^c, K_v^x, K_v^c$ are expected to be rather small while ensuring accuracy of the representations.

2. **Online stage.** For any query parameter $\boldsymbol{\theta}_q$ in the parameter domain:

   (a) Estimate the displacement field $\{u\}(\{X\}, t, \boldsymbol{\theta}_q)$ from expression (11). This requires an interpolation process at $\boldsymbol{\theta} = \boldsymbol{\theta}_q$. For that, we decide to use a diffuse approximation technique [18] that can be used for any parameter space dimension;

   (b) From the estimated displacement field $\{u\}(\{X\}, t^i, \boldsymbol{\theta}_q)$ computed at different instants $t^i \in [0, T_f]$, compute a low-order reduced basis $\{\varphi^k\}(\boldsymbol{\theta}_q), k = 1, \ldots, m_u$ by singular value decomposition. We then get the low-order representations of both displacements and velocities:

   $$\{u\}(\{X\}, t, \boldsymbol{\theta}_q) = \sum_{k=1}^{m_u} \alpha_k(t) \{\varphi^k\}(\boldsymbol{\theta}_q), \qquad (13)$$

   $$\{v\}(\{X\}, t, \boldsymbol{\theta}_q) = \sum_{k=1}^{m_v} \zeta_k(t) \{\gamma^k\}(\boldsymbol{\theta}_q), \qquad (14)$$

   (c) Manifold learning online stage: using diffuse approximation, we determine the low-order manifold $\mathcal{M}$ that links displacements and velocities in the (reduced-order) state space:

   $$\zeta = \mathcal{M}(\boldsymbol{\alpha}, \boldsymbol{\theta}_q);$$

   (d) Derivation of a low-order dynamical system: we then derive a lightweight differential-algebraic dynamical system, easy to solve numerically: for $\boldsymbol{\theta} = \boldsymbol{\theta}_q$, solve

   $$\frac{d\boldsymbol{\alpha}}{dt} = Q\,\zeta(t),$$
   $$\zeta(t) = \mathcal{M}(\boldsymbol{\alpha}(t), \boldsymbol{\theta}_q).$$

   The high-dimensional displacement and velocity fields can then be reconstructed according to (13) and (14).

In the next section, we give all the details of the ROM methodology.

3.2. Offline Stage

3.2.1. Global Parametric Reduced Basis (GPRB)

This first step consists in computing a parametric reduced basis in the whole parameter domain from the database of FOM results (see Section 2.3). For simplification reasons, we use the subscript ϱ that can be either u or v to express displacements and velocity respectively in the formulas.

Let $\mathcal{S}_u^i \in \mathcal{M}_{3N_x, N_c}(\mathbb{R})$ be the matrix of capsule displacement fields $\{u\}$ and $\mathcal{S}_v^i \in \mathcal{M}_{3N_x, N_c}(\mathbb{R})$ the matrix of the velocity fields $\{v\}$ at time t^i, $i = 1, \ldots, N_t$ (Figure 4a), considering all the configurations $\boldsymbol{\theta}_j$ for $j = 1, \ldots, N_c$ of the database, i.e.,

$$\mathcal{S}_u^i = \left[\{u\}(\{X\}, t^i, \boldsymbol{\theta}_1), \ldots, \{u\}(\{X\}, t^i, \boldsymbol{\theta}_{N_c}) \right],$$

and

$$\mathcal{S}_v^i = \left[\{v\}(\{X\}, t^i, \boldsymbol{\theta}_1), \ldots, \{v\}(\{X\}, t^i, \boldsymbol{\theta}_{N_c}) \right].$$

Then we stack all the matrices $\mathcal{S}_\varrho^i$ for $i = 1, \ldots, N_t$ into a big matrix $\mathcal{S}_\varrho \in \mathcal{M}_{3N_x \times N_t, N_c}(\mathbb{R})$:

$$\mathcal{S}_\varrho = \begin{bmatrix} \mathcal{S}_\varrho^1 \\ \mathcal{S}_\varrho^2 \\ \vdots \\ \mathcal{S}_\varrho^{N_t} \end{bmatrix} \text{ for } \varrho = u, v.$$

We then apply SVD [42] and get:

$$\mathcal{S}_\varrho = U_\varrho \Sigma_{\mathcal{S}_\varrho} \Psi_\varrho^T, \text{ for } \varrho = u, v, \tag{15}$$

where $U_\varrho \in \mathcal{M}_{3N_x N_t, N_c}(\mathbb{R})$, $\Psi_\varrho \in \mathcal{M}_{N_c}(\mathbb{R})$ are semi-orthogonal and orthogonal matrices, respectively, and $\Sigma_{\mathcal{S}_\varrho} \in \mathcal{M}_{N_c}(\mathbb{R})$ is the diagonal singular value matrix. The matrix Ψ_ϱ of discrete parameter modes can be truncated according to K_ϱ^c parameters, so we note:

$$\begin{cases} \Psi_\varrho^r = \left[(\Psi_\varrho)_1, \ldots, (\Psi_\varrho)_{K_\varrho^c} \right] \in \mathcal{M}_{N_c, K_\varrho^c}(\mathbb{R}), \\ \text{with } (\Psi_\varrho)_k \in \mathcal{M}_{N_c, 1}(\mathbb{R}) \text{ for } k = 1, \ldots, K_\varrho^c, \text{ and } \varrho = u, v. \end{cases} \tag{16}$$

The orthogonality property ensures that $\left(\Psi_\varrho^r \right)^T \Psi_\varrho^r = I_{K_\varrho^c}$.

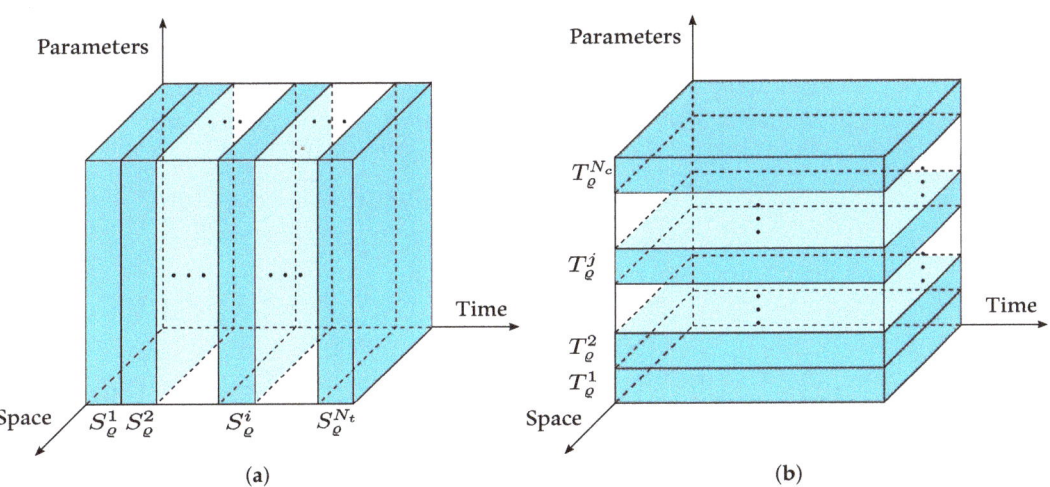

Figure 4. FOM data rearrangements for (**a**) parametric data set selection and (**b**) spatial data set selection.

3.2.2. Global Spatial Reduced Basis (GSRB)

Similarly, we build a global spatial reduced basis that captures the spatial data of capsule displacements. Let $T_u^j \in \mathcal{M}_{3N_x, N_t}(\mathbb{R})$ be the displacement matrix and T_v^j the velocity matrix for the j-th configuration $\boldsymbol{\theta}_j$, for $j = 1, \ldots, N_c$ at all time instants t^i, $i = 1, \ldots, N_t$ (Figure 4b):

$$T_u^j = [\{u\}(\{X\}, t^1, \boldsymbol{\theta}_j), \ldots, \{u\}(\{X\}, t^{N_t}, \boldsymbol{\theta}_j)],$$

and

$$T_v^j = [\{v\}(\{X\}, t^1, \boldsymbol{\theta}_j), \ldots, \{v\}(\{X\}, t^{N_t}, \boldsymbol{\theta}_j)].$$

Then we define the global matrix $\mathcal{T}_\varrho \in \mathcal{M}_{3N_x, N_t \times N_c}(\mathbb{R})$ that horizontally gathers all the matrices T_ϱ^j for $j = 1, \ldots, N_c$ and $\varrho = u, v$, respectively:

$$\mathcal{T}_\varrho = \left[T_\varrho^1, T_\varrho^2, \ldots, T_\varrho^{N_c}\right], \text{ for } \varrho = u, v.$$

The SVD decomposition is applied on $\mathcal{T}_\varrho$ to get

$$\mathcal{T}_\varrho = \Phi_\varrho \Sigma_{T_\varrho} V_\varrho^T, \text{ for } \varrho = u, v, \quad (17)$$

where $\Phi_\varrho \in \mathcal{M}_{3N_x}(\mathbb{R})$, $V_\varrho \in \mathcal{M}_{N_c N_t, 3N_x}(\mathbb{R})$ are orthogonal and semi-orthogonal matrices, respectively, and $\Sigma_{S_\varrho} \in \mathcal{M}_{3N_x}(\mathbb{R})$ is the diagonal singular value matrix with singular values organized in decreasing order. We can also apply a spatial basis truncation at a range of K_ϱ^x for a specified accuracy threshold. The reduced spatial POD basis is stored in the matrix:

$$\Phi_\varrho^r = \left[\{\phi_\varrho\}_1, \ldots, \{\phi_\varrho\}_{K_\varrho^x}\right] \in \mathcal{M}_{3N_x, K_\varrho^x}(\mathbb{R}) \quad (18)$$

with the orthogonality property $\left(\Phi_\varrho^r\right)^T \Phi_\varrho^r = I_{K_\varrho^x}$, $\varrho = u, v$.

3.3. Data Dimensionality Reduction

Once the POD modes of $\mathcal{S}_\varrho$ and $\mathcal{T}_\varrho$ for the displacement fields ($\varrho = u$) and the velocity fields ($\varrho = v$) are computed, one can summarize (approximate) capsule displacement and velocity fields of the database at any discrete time t^i ($i = 1 \ldots, N_t$) as

$$\{u\}\left(\{X\}, t^i, [\boldsymbol{\theta}_1, \ldots, \boldsymbol{\theta}_{N_c}]\right) \approx \Phi_u^r A(t^i) (\Psi_u^r)^T \in \mathcal{M}_{3N_x, N_c}(\mathbb{R}), \quad (19)$$

$$\{v\}\left(\{X\}, t^i, [\boldsymbol{\theta}_1, \ldots, \boldsymbol{\theta}_{N_c}]\right) \approx \Phi_v^r B(t^i) (\Psi_v^r)^T \in \mathcal{M}_{3N_x, N_c}(\mathbb{R}), \quad (20)$$

where $A(t^i) \in \mathcal{M}_{K_u^x, K_u^c}(\mathbb{R})$ and $B(t^i) \in \mathcal{M}_{K_v^x, K_v^c}(\mathbb{R})$ are some coefficient matrices depending on time t^i. If the approximation is chosen as the orthogonal projection over the vector spaces spanned by the POD modes, the coefficient matrices are computed as follows for $i = 1 \ldots, N_t$:

$$A(t^i) = \underbrace{(\Phi_u^r)^T}_{K_u^x \times (3N_x)} \underbrace{\{u\}\left(\{X\}, t^i, [\boldsymbol{\theta}_1, \ldots, \boldsymbol{\theta}_{N_c}]\right)}_{(3N_x) \times N_c} \underbrace{\Psi_u^r}_{N_c \times K_u^c}, \quad (21)$$

$$B(t^i) = \underbrace{(\Phi_v^r)^T}_{K_v^x \times (3N_x)} \underbrace{\{v\}\left(\{X\}, t^i, [\boldsymbol{\theta}_1, \ldots, \boldsymbol{\theta}_{N_c}]\right)}_{(3N_x) \times N_c} \underbrace{\Psi_v^r}_{N_c \times K_v^c}. \quad (22)$$

The outputs of the offline stage are respectively the POD matrices $\Phi_u^r, \Phi_v^r, \Psi_u^r, \Psi_v^r$ and the small matrices $A(t^i), B(t^i)$, $i = 1, \ldots, N_t$. The next online stage will operate on the summarized data (19), (20) with coefficients matrices (21), (22). The algorithm of the offline phase is summarized in Algorithm 1.

Algorithm 1 Offline phase

Require: database of $\boldsymbol{\theta}_k$ for $k = 1, \ldots, N_c$, truncations K_ϱ^c, number of snapshots N_t.
    for $i \leftarrow 1, \ldots, N_t$ do
        if $(\varrho = u)$ then
            $\mathcal{S}_u^i \leftarrow \left[\{u\}(\{X\}, t^i, \boldsymbol{\theta}_1), \ldots, \{u\}(\{X\}, t^i, \boldsymbol{\theta}_{N_c}) \right]; \mathcal{S}_u \leftarrow [\mathcal{S}_u; \mathcal{S}_u^i];$
        else
            $\mathcal{S}_v^i \leftarrow \left[\{v\}(\{X\}, t^i, \boldsymbol{\theta}_1), \ldots, \{v\}(\{X\}, t^i, \boldsymbol{\theta}_{N_c}) \right]; \mathcal{S}_v \leftarrow [\mathcal{S}_v; \mathcal{S}_v^i];$
        end if
    end for
    for $j \leftarrow 1, \ldots, N_c$ do
        if $(\varrho = u)$ then
            $\mathcal{T}_u^j \leftarrow \left[\{u\}(\{X\}, t^1, \boldsymbol{\theta}_j), \ldots, \{u\}(\{X\}, t^{N_t}, \boldsymbol{\theta}_j) \right]; \mathcal{T}_u \leftarrow [\mathcal{T}_u, \mathcal{T}_u^j];$
        else
            $\mathcal{T}_v^j \leftarrow \left[\{v\}(\{X\}, t^1, \boldsymbol{\theta}_j), \ldots, \{v\}(\{X\}, t^{N_t}, \boldsymbol{\theta}_j) \right]; \mathcal{T}_v \leftarrow [\mathcal{T}_v, \mathcal{T}_v^j];$
        end if
    end for
    $\Phi_\varrho \leftarrow \text{SVD}(\mathcal{S}_\varrho), \Psi_\varrho \leftarrow \text{SVD}(\mathcal{T}_\varrho)$, for $\varrho \leftarrow u, v$;
    for $i = 1, \ldots, N_t$ do
        $A(t^i) \leftarrow (\Phi_u^r)^T \{u\}(\{X\}, t^i, [\boldsymbol{\theta}_1, \ldots, \boldsymbol{\theta}_{N_c}]) \Psi_u^r;$
        $B(t^i) \leftarrow (\Phi_v^r)^T \{v\}(\{X\}, t^i, [\boldsymbol{\theta}_1, \ldots, \boldsymbol{\theta}_{N_c}]) \Psi_v^r;$
    end for

3.4. Online Stage: Search for an Approximate Solution at a Query Configuration $\boldsymbol{\theta}_q$

In the online stage, a user will ask for an approximate solution at a new (query) configuration $\boldsymbol{\theta} = \boldsymbol{\theta}_q$ that has not been already computed by the FOM solver or is not stored in the database. Ingredients of the online stage will be: (i) the data summarization of the previous offline stage; (ii) a first estimation of the spatio-temporal solution at $\boldsymbol{\theta} = \boldsymbol{\theta}_q$; (iii) the computation of a low-dimensional spatial reduced basis suitable for $\boldsymbol{\theta} = \boldsymbol{\theta}_q$; (iv) the construction of a manifold $\mathcal{M}$ that links variables of displacements and velocities in the low-order state space to solve the equation of membrane mechanics; (v) finally, the building of a low-order differential-algebraic (DAE) system of equations that defines the reduced-order model. Substeps (ii) and (iv) will make use of diffuse approximation (DA) as a universal approximator for multivariate functions.

3.4.1. First Estimation of the Solutions at $\boldsymbol{\theta} = \boldsymbol{\theta}_q$

As an introduction, let us assume that, from the parameter sampling $\{\boldsymbol{\theta}_1, \ldots, \boldsymbol{\theta}_{N_c}\}$, we consider a polynomial Lagrange interpolation with Lagrange polynomials denoted by $L_i(\boldsymbol{\theta})$ such that the Lagrange property

$$L_i(\boldsymbol{\theta}_j) = \delta_{ij}, \quad 1 \leq i, j \leq N_c$$

is fulfilled (δ_{ij} is the standard Kronecker symbol). Let us denote by $L(\boldsymbol{\theta}) = (L_j(\boldsymbol{\theta}))_{j=1,\ldots,N_c} \in \mathbb{R}^{N_c}$ the vector that stores the Lagrange polynomials. Then

$$\mathcal{I}\{u\}\left(\{X\}, t^i, \boldsymbol{\theta}_q\right) := \{u\}\left(\{X\}, t^i, [\boldsymbol{\theta}_1, \ldots, \boldsymbol{\theta}_{N_c}]\right) L(\boldsymbol{\theta}_q) \in \mathbb{R}^{3N_x}$$

is an interpolated displacement field at parameter $\boldsymbol{\theta} = \boldsymbol{\theta}_q$ and discrete time $t = t^i$. One can of course do the same for the velocity field.

Unfortunately, Lagrange polynomial interpolation is not suitable for parameter spaces of arbitrary dimension because of the curse of dimensionality and because it may suffer from instability issues (Runge phenomenon). Rather than using polynomial interpolation, we propose to use a Diffuse Approximation (DA) technique [18,29] which is an approximation method based on local low-order polynomial reconstruction (of order one or two) using a compactly-supported kernel function and weighted least squares. The DA method is known to be a robust and reliable approach which is less sensitive to the location of the sampling points. Moreover, it can be applied to multivariate functions of arbitrary dimensions, which is interesting for larger or more general parameter spaces. It is particularly suited for the current problem, for which the sampling is performed on a Cartesian grid. It may fail in the occurrence of local point alignment within the cloud points, which does not occur in the present study. The accuracy of the DA method may slightly decrease close to the boundary of the domain, the number of neighboring points being reduced.

To estimate the displacement field for $\boldsymbol{\theta} = \boldsymbol{\theta}_q$, we look for a vector $\boldsymbol{\psi}_u(\boldsymbol{\theta}_q) \in \mathbb{R}^{K_u^c}$ such that

$$\{u\}(\{x\}, t^i, \boldsymbol{\theta}_q) = \Phi_u^r A(t^i) \, \boldsymbol{\psi}_u(\boldsymbol{\theta}_q) \tag{23}$$

returns an approximation of the displacement field at $\boldsymbol{\theta} = \boldsymbol{\theta}_q$. Similarly for the velocity field, we search for a vector $\boldsymbol{\psi}_v(\boldsymbol{\theta}_q) \in \mathbb{R}^{K_v^c}$ that gives

$$\{v\}(\{x\}, t^i, \boldsymbol{\theta}_q) = \Phi_v^r B(t^i) \, \boldsymbol{\psi}_v(\boldsymbol{\theta}_q). \tag{24}$$

Each vector $\boldsymbol{\psi}_\varrho(\boldsymbol{\theta}_q) \in \mathbb{R}^{K_\varrho^c}$ can be locally approximated by

$$\boldsymbol{\Psi}_\varrho(\boldsymbol{\theta}_q) = \mathcal{A}_\varrho \, p(\boldsymbol{\theta}_q), \text{ for } \varrho = u, v, \tag{25}$$

where the matrix $\mathcal{A}_\varrho \in \mathcal{M}_{K_\varrho^c, m}(\mathbb{R})$ (to be determined) is the approximation coefficient matrix and $p(\boldsymbol{\theta}_q) \in \mathbb{R}^m$ is a vector of independent polynomial functions, where

$$\begin{cases} p(\boldsymbol{\theta}) = \begin{bmatrix} 1 & Ca & a/\ell \end{bmatrix}^T, \ m=3 & \text{for first order DA,} \\ p(\boldsymbol{\theta}) = \begin{bmatrix} 1 & Ca & a/\ell & Ca(a/\ell) & (Ca)^2 & (a/\ell)^2 \end{bmatrix}^T, \ m=6 & \text{for second order DA.} \end{cases} \tag{26}$$

To approximate $\boldsymbol{\psi}_\varrho(\boldsymbol{\theta}_q)$, let us consider a neighborhood $\mathscr{S}(\boldsymbol{\theta}_q)$ centered on $\boldsymbol{\theta}_q$ containing M neighboring points (Figure 5a). It is an ellipse of equation

$$(\theta_1 - (\boldsymbol{\theta}_q)_1)^2 + \tilde{r}^2(\theta_2 - (\boldsymbol{\theta}_q)_2)^2 = R^2$$

where $\tilde{r}$ is fixed (equal to 1.9 in Figure 5a) and R is chosen such that the ellipse contains M points (M being chosen by the operator). In other words, the distance between $\boldsymbol{\theta} = (\theta_1, \theta_2)^T$ and $\boldsymbol{\theta}_q$ is

$$d = \left((\theta_1 - (\boldsymbol{\theta}_q)_1)^2 + \tilde{r}^2(\theta_2 - (\boldsymbol{\theta}_q)_2)^2\right)^{\frac{1}{2}} / R. \tag{27}$$

The compactly supported Wendland weight function shown in Figure 5b is classically used. It has appropriate high-order approximation properties ([43]):

$$\begin{cases} w(d) = 2\,d^3 - 3\,d^2 + 1, & d \leq 1, \\ 0, & \text{otherwise.} \end{cases} \tag{28}$$

Diffuse approximation consists in minimizing the weighted least square problem

$$\min_{\mathcal{A}_\varrho \in \mathcal{M}_{K_\varrho^c, m}(\mathbb{R})} J_{\boldsymbol{\theta}_q}(\mathcal{A}_\varrho) := \frac{1}{2} \sum_{\boldsymbol{\theta} \in \mathscr{S}(\boldsymbol{\theta}_q)} w(d(\boldsymbol{\theta})) \left\| \mathcal{A}_\varrho \, p(\boldsymbol{\theta}) - [\boldsymbol{\Psi}_\varrho^r(\boldsymbol{\theta})]^T \right\|_{\mathbb{R}^{K_\varrho^c}}^2 \tag{29}$$

where $[\boldsymbol{\Psi}_\varrho^r(\boldsymbol{\theta})]^T$ is the truncated matrix of modes that correspond to couples $\boldsymbol{\theta}_k$, $k = 1, \ldots, N_c$. The solution $\mathcal{A}_\varrho$ ($\varrho = u, v$) of the weighted least square problem (29) is then

$$\mathcal{A}_\varrho = (\boldsymbol{\Psi}_\varrho^r)^T \mathcal{WP} \left(\mathcal{P}^T \mathcal{WP}\right)^{-1} \in \mathcal{M}_{K_\varrho^c, m}(\mathbb{R}) \tag{30}$$

where the matrix $\mathcal{P} \in \mathcal{M}_{N_c,m}(\mathbb{R})$ and the diagonal matrix of weights $\mathcal{W} \in \mathcal{M}_{N_c}(\mathbb{R})$ are defined as

$$\mathcal{P} = \begin{bmatrix} p(\boldsymbol{\theta}_1)^T \\ \vdots \\ p(\boldsymbol{\theta}_{N_c})^T \end{bmatrix} \text{ and } \mathcal{W} = \begin{bmatrix} w_1 & 0 & \cdots & 0 \\ 0 & w_2 & & \vdots \\ \vdots & & \ddots & \\ 0 & \cdots & & w_{N_c} \end{bmatrix}. \quad (31)$$

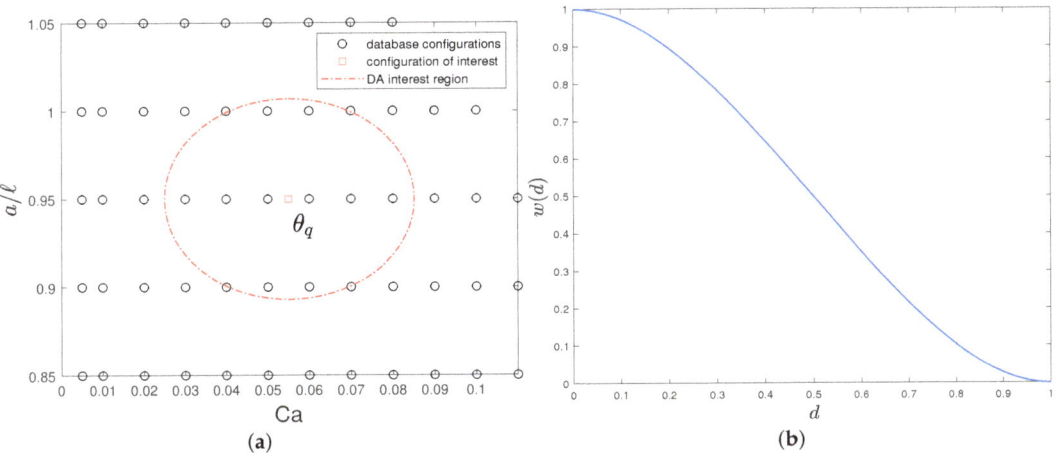

Figure 5. (a) DA elliptical region of interest (dashed line) defined around the point $\boldsymbol{\theta}_q = (Ca = 0.055, a/\ell = 0.95)$ in the parametric space with $M = 10$ neighbors; (b) Weight function $w(d)$.

3.4.2. Construction of a Low-Order Reduced Basis Suitable for $\boldsymbol{\theta} = \boldsymbol{\theta}_q$, Data Generation

From (23) and (24), one can easily generate some pseudo-snapshot matrices $\mathcal{U}(\boldsymbol{\theta}_q)$ and $\mathcal{V}(\boldsymbol{\theta}_q)$ that gather the estimated displacements and velocities at N_t discrete times, respectively:

$$\begin{cases} \mathcal{U}(\boldsymbol{\theta}_q) = \left[\{u\}(\{X\}, t^1, \boldsymbol{\theta}_q), \ldots, \{u\}(\{X\}, t^{N_t}, \boldsymbol{\theta}_q) \right], \\ \mathcal{V}(\boldsymbol{\theta}_q) = \left[\{v\}(\{X\}, t^1, \boldsymbol{\theta}_q), \ldots, \{v\}(\{X\}, t^{N_t}, \boldsymbol{\theta}_q) \right]. \end{cases} \quad (32)$$

One can then apply a new SVD decomposition of matrices $\mathcal{U}(\boldsymbol{\theta}_q)$ and $\mathcal{V}(\boldsymbol{\theta}_q)$ respectively to get spatial POD modes $\{\boldsymbol{\varphi}^k\}(\boldsymbol{\theta}_q) \in \mathbb{R}^{3N_x}, k = 1, \ldots, m_u$ for $\{u\}$ and velocity POD modes $\{\boldsymbol{\gamma}^k\}(\boldsymbol{\theta}_q) \in \mathbb{R}^{3N_x}, k = 1, \ldots, m_v$ for $\{v\}$.

$$\text{POD}(\mathcal{U}(\boldsymbol{\theta}_q)) \rightarrow \{\boldsymbol{\varphi}^1\}(\boldsymbol{\theta}_q), \ldots, \{\boldsymbol{\varphi}^{m_u}\}(\boldsymbol{\theta}_q) \quad (33)$$

$$\text{POD}(\mathcal{V}(\boldsymbol{\theta}_q)) \rightarrow \{\boldsymbol{\gamma}^1\}(\boldsymbol{\theta}_q), \ldots, \{\boldsymbol{\gamma}^{m_v}\}(\boldsymbol{\theta}_q) \quad (34)$$

where m_u and m_v are the truncation ranks of displacement and velocity modes determined in the next section on numerical experiments. One can then search the displacement and velocity fields at $\boldsymbol{\theta} = \boldsymbol{\theta}_q$ as

$$\{u\}(\{X\}, t, \boldsymbol{\theta}_q) = \sum_{k=1}^{m_u} \alpha_k(t) \{\boldsymbol{\varphi}^k\}(\boldsymbol{\theta}_q), \quad (35)$$

$$\{v\}(\{X\}, t, \boldsymbol{\theta}_q) = \sum_{k=1}^{m_v} \xi_k(t) \{\boldsymbol{\gamma}^k\}(\boldsymbol{\theta}_q). \quad (36)$$

By denoting

$$\Phi(\boldsymbol{\theta}_q) = \left[\{\boldsymbol{\varphi}^1\}(\boldsymbol{\theta}_q), \ldots, \{\boldsymbol{\varphi}^{m_u}\}(\boldsymbol{\theta}_q) \right] \in \mathcal{M}_{3N_x, m_u}(\mathbb{R}), \tag{37}$$

$$\Gamma(\boldsymbol{\theta}_q) = \left[\{\boldsymbol{\gamma}^1\}(\boldsymbol{\theta}_q), \ldots, \{\boldsymbol{\gamma}^{m_v}\}(\boldsymbol{\theta}_q) \right] \in \mathcal{M}_{3N_x, m_v}(\mathbb{R}) \tag{38}$$

and $\boldsymbol{\alpha}(t) = [\alpha_1(t), \ldots, \alpha_{m_u}(t)]^T \in \mathbb{R}^{m_u}$, $\boldsymbol{\xi}(t) = [\xi_1(t), \ldots, \xi_{m_v}(t)]^T \in \mathbb{R}^{m_v}$, we have the vector formulas

$$\{u\}(\{X\}, t, \boldsymbol{\theta}_q) = \Phi(\boldsymbol{\theta}_q)\,\boldsymbol{\alpha}(t), \quad \{v\}(\{X\}, t, \boldsymbol{\theta}_q) = \Gamma(\boldsymbol{\theta}_q)\,\boldsymbol{\xi}(t). \tag{39}$$

The mode matrices $\Phi(\boldsymbol{\theta}_q)$ and $\Gamma(\boldsymbol{\theta}_q)$ are assumed to be orthonormal (w.r.t the natural Euclidean inner product), so we have $[\Phi(\boldsymbol{\theta}_q)]^T \Phi(\boldsymbol{\theta}_q) = I_{m_u}$ and $[\Gamma(\boldsymbol{\theta}_q)]^T \Gamma(\boldsymbol{\theta}_q) = I_{m_v}$.

3.4.3. Toward a Physically Consistent Dynamical Reduced-Order Model

Consider now the forward Euler scheme on the FSI system with a ROM time step $\delta t^{ROM} > 0$: at time $t^{i+1,ROM} = t^{i,ROM} + \delta t^{ROM}$, the numerical scheme is

$$\{u^{i+1}\} = \{u^i\} + \delta t^{ROM} \{v^i\}, \tag{40}$$

$$\{v^{i+1}\} = \{\mathcal{N}\}(\{u^{i+1}\}, \boldsymbol{\theta}_q). \tag{41}$$

Let us emphasize that the equation of local mechanical equilibrium depends on the parameter $\boldsymbol{\theta}_q$. For the reduced-order model, we would like to have a similar algebraic structure to (40), (41) but formulated as a low-dimensional system. If $\{u^i\}$ and $\{v^i\}$ are searched in the form $\{u^i\} = \Phi(\boldsymbol{\theta}_q)\,\boldsymbol{\alpha}^i$ and $\{v^i\} = \Gamma(\boldsymbol{\theta}_q)\,\boldsymbol{\xi}^i$, respectively, Equation (40) becomes

$$\Phi(\boldsymbol{\theta}_q)\,\boldsymbol{\alpha}^{i+1} = \Phi(\boldsymbol{\theta}_q)\,\boldsymbol{\alpha}^i + \delta t^{ROM} \Gamma(\boldsymbol{\theta}_q)\,\boldsymbol{\xi}^i.$$

By multiplying by $[\Phi(\boldsymbol{\theta}_q)]^T$ on the left, we get the system of m_u equations

$$\boldsymbol{\alpha}^{i+1} = \boldsymbol{\alpha}^i + \delta t^{ROM} Q(\boldsymbol{\theta}_q)\,\boldsymbol{\xi}^i, \tag{42}$$

where $Q(\boldsymbol{\theta}_q) = [\Phi(\boldsymbol{\theta}_q)]^T \Gamma(\boldsymbol{\theta}_q)$. Equation (41) is replaced by

$$\Gamma(\boldsymbol{\theta}_q)\,\boldsymbol{\xi}^{i+1} = \{\mathcal{N}\}(\Phi(\boldsymbol{\theta}_q)\,\boldsymbol{\alpha}^{i+1}, \boldsymbol{\theta}_q).$$

By multiplying by $[\Gamma(\boldsymbol{\theta}_q)]^T$ on the left, we get

$$\boldsymbol{\xi}^{i+1} = \mathcal{M}(\boldsymbol{\alpha}^{i+1}, \boldsymbol{\theta}_q)$$

where

$$\mathcal{M}(\boldsymbol{\alpha}^{i+1}, \boldsymbol{\theta}_q) = [\Gamma(\boldsymbol{\theta}_q)]^T \{\mathcal{N}\}(\Phi(\boldsymbol{\theta}_q)\,\boldsymbol{\alpha}^{i+1}, \boldsymbol{\theta}_q) \in \mathbb{R}^{m_v}. \tag{43}$$

3.4.4. Manifold Learning

Because of nonlinear terms, the direct computation of $\mathcal{M}(\boldsymbol{\alpha}^{i+1}, \boldsymbol{\theta}_q)$ in (43) requires high-dimensional computations, which makes the ROM irrelevant from a performance point of view. To "identify" a low-order manifold $\mathcal{M}$, we rather adopt a data-driven approach based once again of diffuse approximation. We link the entry data $\alpha_k^D(t^i)$, $k = 1, \ldots, m_u$, $i = 1, \ldots, N_t$ to the output data $\xi_k^D(t^i)$, $k = 1, \ldots, m_v$, $i = 1, \ldots, N_t$ ('D' stands for 'data'). For that, one can compute the orthogonal projections of the pseudo-snapshots over the POD bases, leading to the formulas

$$\alpha_k^D(t^i) = \langle \{u\}(\{X\}, t^i, \boldsymbol{\theta}_q), \{\boldsymbol{\varphi}^k\}(\boldsymbol{\theta}_q) \rangle$$

and

$$\xi^D(t^i) = \langle \{v\}(\{X\}, t^i, \boldsymbol{\theta}_q), \{\boldsymbol{\gamma}^k\}(\boldsymbol{\theta}_q) \rangle$$

at instants $t^i = i\Delta t$. Manifold learning consists in achieving a (nonlinear) regression method that links entry and output data. We are looking for a manifold representation $\xi = \mathcal{M}(\alpha, \theta_q)$ in the form

$$\xi_k = p(\alpha)^T a^k, \quad k = 1, \ldots, m_v \tag{44}$$

where $p(\alpha)$ is the vector made of monomials in α of order zero and one, and $a^k \in \mathbb{R}^{m_u+1}$ is a vector of coefficients to be determined from the data. This corresponds to a local linear embedding process. For each $k = 1, \ldots, m_v$, one looks for a coefficient vector $a^k(t) \in \mathbb{R}^{m_u+1}$ solution of the weighted least square problem

$$a^k(t) = \arg\min_{a \in \mathbb{R}^{m_u+1}} \frac{1}{2} \sum_{i=1}^{N_t} w\left(\frac{|t - t^i|}{R}\right) \left(p(\alpha^D(t^i))^T a - \xi_k^D(t^i)\right)^2 \tag{45}$$

where $t \in [0, T_f]$, $w = w(d)$ is the weight function defined in Figure 5b and $d = \frac{|t-t^i|}{R}$. This returns a regression function

$$\xi_k = \xi_k(t, \alpha(t)) = p(\alpha(t))^T a^k(t). \tag{46}$$

3.4.5. Low-Order Dynamical Reduced Order Model

The resulting time-discrete reduced-order model is then

$$t^{i+1, ROM} = t^{i, ROM} + \delta t^{ROM}, \tag{47}$$

$$\alpha^{i+1} = \alpha^i + \delta t^{ROM} Q(\theta_q) \xi^i, \tag{48}$$

$$\xi_k^{i+1} = p(\alpha^{i+1})^T a^k(t^{i+1, ROM}) \quad \forall k \in \{1, \ldots, m_v\}. \tag{49}$$

High-dimensional displacement and velocity fields can be reconstructed as follows:

$$\{u\}\left(\{X\}, t^{i+1, ROM}, \theta_q\right) = \Phi(\theta_q) \alpha^{i+1}, \quad \{v\}\left(\{X\}, t^{i+1, ROM}, \theta_q\right) = \Gamma(\theta_q) \xi^{i+1}.$$

The online stage of the reduced-order model is summarized in Algorithm 2.

Algorithm 2 Online phase

Require: choose a query parameter θ_q, choose a time step $\delta t^{ROM} > 0$.
  Initialization: $t = t^{0, ROM} = 0$, $\alpha^0 = 0$, $\xi^0 = \xi^D(0)$;
  Compute $\Psi_u(\theta_q)$ and $\Psi_v(\theta_q)$ from the diffuse approximation approach;
  for $i = 1 \ldots, N_t$ **do**
    $\{u\}(\{x\}, t^i, \theta_q) \leftarrow \Phi_u^r A(t^i) \Psi_u(\theta_q)$;
    $\{v\}(\{x\}, t^i, \theta_q) \leftarrow \Phi_v^r B(t^i) \Psi_v(\theta_q)$;
  end for
  $\mathcal{U}(\theta_q) \leftarrow [\{u\}(\{x\}, t^1, \theta_q), \ldots, \{u\}(\{x\}, t^{N_t}, \theta_q)]$;
  $\mathcal{V}(\theta_q) \leftarrow [\{v\}(\{x\}, t^1, \theta_q), \ldots, \{v\}(\{x\}, t^{N_t}, \theta_q)]$;
  Compute $\Phi(\theta_q)$, $\Gamma(\theta_q)$, $Q(\theta_q)$, $\alpha^D(t^i)$ and $\xi^D(\theta_i), i = 1, \ldots, N_t$;
  while $t < T_f$ **do**
    $t \leftarrow t + \delta t^{ROM}$; $t^{i+1, ROM} = t^{i, ROM} + \delta t^{ROM}$;
    $\alpha^{i+1} = \alpha^i + \delta t^{ROM} Q(\theta_q) \xi^i$;
    Compute $a^k(t^{i+1, ROM})$, $k = 1, \ldots, m_v$ from the diffuse approximation approach;
    $\xi_k^{i+1} = p(\alpha^{i+1})^T a^k(t^{i+1, ROM})$;
    If needed, reconstruct the high-dimensional displacements/velocity fields:
    $\{u\}(\{X\}, t^{i+1, ROM}, \theta_q) = \Phi(\theta_q) \alpha^{i+1}$;
    $\{v\}(\{X\}, t^{i+1, ROM}, \theta_q) = \Gamma(\theta_q) \xi^{i+1}$;
  end while

4. Numerical Experiments

4.1. Study Case

We consider a capsule flowing in a square-base microchannel of base edges of length 2ℓ. We want to capture the capsule dynamics for capillary numbers Ca belonging to the interval $[0.005, 0.2]$ and aspect ratios a/ℓ in the interval $[0.75, 1.2]$ for which a steady state shape is reached. The Caps3D code [38,40] is then used as FOM solver. The comparison of the FOM results with experimental ones using a square-base cylinder have been thoroughly described in other previous studies (see for example [23,24,27,39] from A.V. Salsac's research team). The total non-dimensional time for simulation is $T = 20$. For any capillary number and aspect ratio, the capsule is discretized with the same mesh resolution and connectivity, consisting of $N_x = 2562$ nodes (corresponding to 1280 triangular elements), with a capsule mesh size $\Delta h_C = 0.075\, a$ (see Figure 6). A second-order RK2 Ralston scheme is used for time integration. The dimensionless time step is $\dot{\gamma}\delta t = 5 \cdot 10^{-4}$ for $Ca > 0.02$ and $\dot{\gamma}\delta t = 10^{-4}$ for $Ca \leq 0.02$.

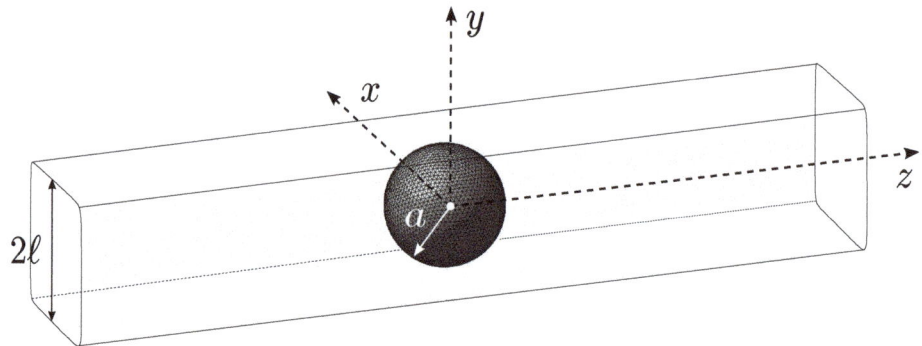

Figure 6. Three-dimensional representation of a capsule flowing in a square microchannel at $T = 0$.

4.2. FOM Result Database Generation

A database of FOM results is generated from a sampling of the parameter domain (see Figure 7). It is observed that configurations for which a shape steady state is reached before the non-dimensional final time of 20 correspond to couples $(Ca, a/\ell)$ in the parameter plane below the dashed red line of Figure 7. Using a Cartesian parameter sampling with step sizes of 0.01 in Ca and 0.05 in a/ℓ, plus few additional points at $Ca = 0.005$, we get a database made of $N_c = 118$ configurations. From Caps3D FOM solutions, we pick up time-snapshot solutions every time step $\Delta t = 0.2$ in non-dimensional time scale, corresponding to $N_t = 100$. This makes a datacube made of $2 \times 3 N_x N_c N_t \approx 1.81 \cdot 10^8$ double precision float numbers taking about 1.45 GB of memory.

Clustering Strategy

For the sake of memory storage complexity, we adopt a strategy of data clustering with two weakly-overlapping clusters chosen manually, represented in Figure 7. For each cluster, a data dimensionality reduction is done following the offline-stage algorithm presented in Section 3. That means that two families of reduced-order models are actually computed. In the online stage, for a new query parameter vector $\boldsymbol{\theta}_q$, one has to determine the cluster of belonging.

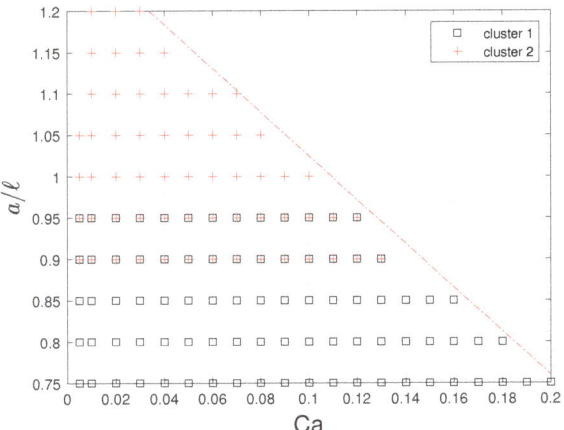

Figure 7. Design of computer experiment with sampling in the admissible parameter domain. The parameter domain is splitted up into two overlapping clusters: cluster 1 (squares), cluster 2 (crosses) and overlapping region (mixed squares and crosses).

4.3. Elements of Analysis—Accuracy Criteria

In order to measure the approximation error generated by the data dimensionality process, we introduce the classical Relative Information Content (RIC) (see for example [9]), which is computed as:

$$\text{RIC}(K) = \frac{\sum_{k=K+1}^{r} \tilde{\sigma}_k^2}{\sum_{k=1}^{r} \tilde{\sigma}_k^2}, \quad K = 1, \ldots, r, \qquad (50)$$

where $\tilde{\sigma}_k$ is the k-th singular value from the SVD decomposition, r is the rank of the matrix of study ($\mathcal{S}_\varrho$ or $\mathcal{T}_\varrho$) and K is the truncation rank. A supplementary indicator is the ratio

$$K \mapsto \frac{\tilde{\sigma}_K}{\tilde{\sigma}_1} \qquad (51)$$

that gives an idea of the decay rate of the singular values.

The second criterion directly measures the error between the shape predicted by the ROM and the shape computed by the FOM. This is achieved by using the so-called Modified Hausdorff distance d_{MH} [44] that we normalize by the capsule radius a. The modified Hausdorff distance computes the distance between two finite sets $\mathcal{F}$ and $\mathcal{G}$ of a normed space of norm $\|.\|$, and is defined as

$$d_{MH}(\mathcal{F}, \mathcal{G}) = \max(d_h(\mathcal{F}, \mathcal{G}), d_h(\mathcal{F}, \mathcal{G})), \qquad (52)$$

with

$$d_h(\mathcal{F}, \mathcal{G}) = \frac{1}{N_\mathcal{F}} \sum_{p_\mathcal{F} \in \mathcal{F}} d_s(p_\mathcal{F}, \mathcal{G}) \qquad (53)$$

where $N_\mathcal{F}$ is the number of points in the set $\mathcal{F}$ and $d_s(p_\mathcal{F}, \mathcal{G})$ is the distance between $p_\mathcal{F}$ and the set $\mathcal{G}$, which is defined as

$$d_s(p_\mathcal{F}, \mathcal{G}) = \min_{p_\mathcal{G} \in \mathcal{G}} \|p_\mathcal{F} - p_\mathcal{G}\|. \qquad (54)$$

4.4. Dimensionality Reduction Analysis

A singular value decomposition analysis is first performed on the matrices $\mathcal{S}_u$ and $\mathcal{S}_v$, and then on $\mathcal{T}_u$ and $\mathcal{T}_v$. In Figure 8a, we plot the indicator $(1-\text{RIC})$ (see (50)), as a function of the truncation rank K, for $\mathcal{S}_u$ and $\mathcal{S}_v$. What can be seen is that $(1-\text{RIC})$ rapidly converges towards the value 0 in all cases. An expected $(1-\text{RIC})$ of 10^{-7} is reached for a truncation rank K_u^c (resp. K_v^c) of 7 for the displacement (resp. 23 for the velocity). Similarly in Figure 8b, we plot the indicator $(1-\text{RIC})$ for $\mathcal{T}_u$ and $\mathcal{T}_v$. The number of modes K_u^x (resp. K_v^x) needed to reach the threshold of 10^{-7} is 7 for the displacement (resp. 56 for the velocity).

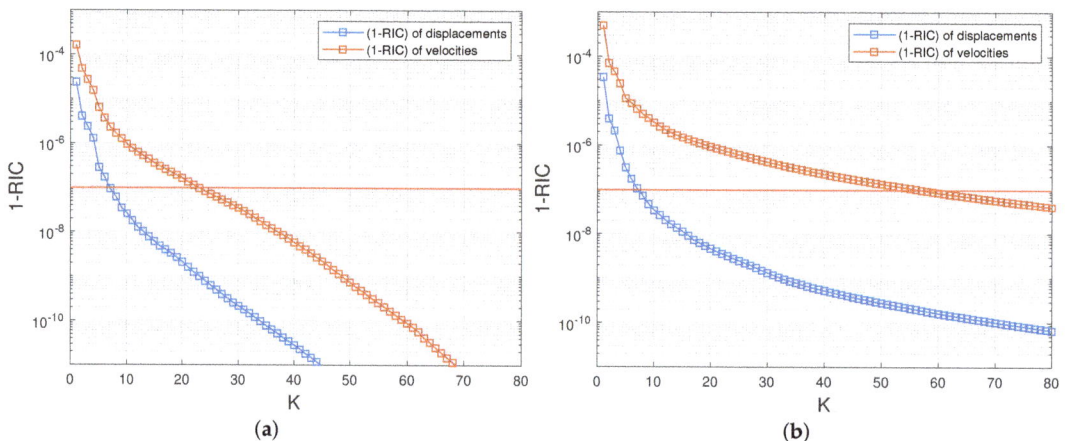

Figure 8. Behaviour of the relative information content of the matrices $\mathcal{S}_u$ and $\mathcal{S}_v$ (**a**) and $\mathcal{T}_u$ and $\mathcal{T}_v$ (**b**) shown in the form $(1-\text{RIC})$ as a function of the truncation rank K. The horizontal red line corresponds to $(1-\text{RIC}) = 10^{-7}$.

As supplementary indicators, the singular values $\tilde{\sigma}_K$ normalized by $\tilde{\sigma}_1$ are plotted in Figure 9a (resp. Figure 9b) for both matrices $\mathcal{S}_u$ and $\mathcal{S}_v$ (resp. $\mathcal{T}_u$ and $\mathcal{T}_v$) in $\log_{10}$ scale. One can first observe a lower decay rate for the velocity fields compared to the displacements, meaning a greater information complexity for the velocity. Secondly, the decay rate is lower for the global spatial mode than for the parametric modes, indicating a larger entropy of information on the whole parameter domain. That justifies the derivation of suitable lower order spatial basis at a query parameter $\boldsymbol{\theta}_q$ in the online stage.

At the beginning of the online stage, for a query parameter $\boldsymbol{\theta}_q$, an interpolated approximate solution is computed thanks to a diffuse approximation reconstruction. This allows us to get pseudo-snapshots in time for both displacements and velocities, stored in matrices $\mathcal{U}(\boldsymbol{\theta}_q)$ and $\mathcal{V}(\boldsymbol{\theta}_q)$, respectively. We assess the RIC for the two matrices, from an experimental parameter vector $\boldsymbol{\theta}_q = (0.10, 0.90)$. The comparison of the time evolution of POD coefficients between FOM and ROM models shows a high accuracy (see Figure 10). and Figure 11 shows that the RIC rapidly converges to 1. An expected RIC greater than $1-10^{-7}$ returns a truncation rank m_u (resp. m_v) of value 3 (resp. 8).

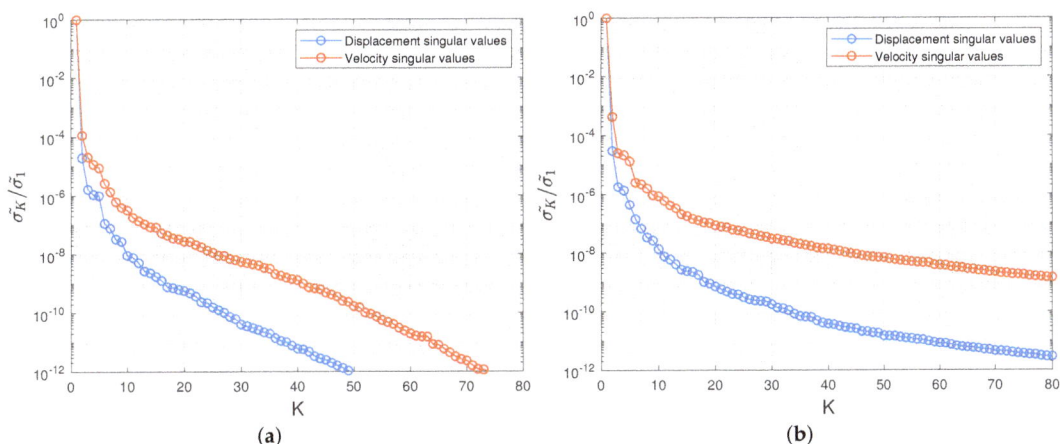

Figure 9. (a) Parametric normalized singular values $\tilde{\sigma}_K/\tilde{\sigma}_1$ for $\mathcal{S}_u$ and $\mathcal{S}_v$; (b) Spatial normalized singular values $\tilde{\sigma}_K/\tilde{\sigma}_1$ for $\mathcal{T}_u$ and $\mathcal{T}_v$.

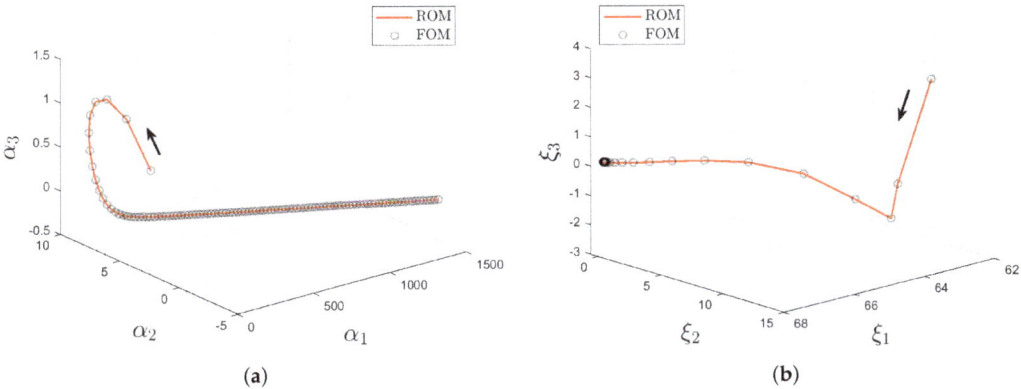

Figure 10. FOM versus ROM comparison of the time evolution of the first three displacement (a) and velocity (b) POD coefficients for the query parameter $\boldsymbol{\theta}_q = (0.10, 0.90)$.

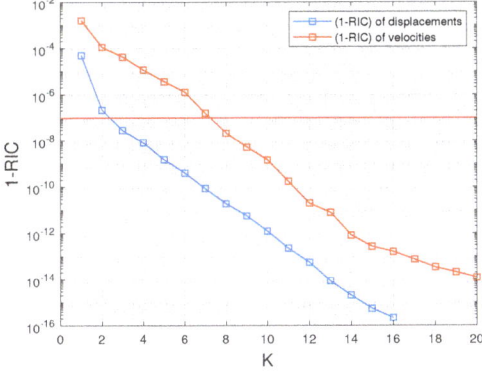

Figure 11. Online stage: behaviour of the relative information content of the matrices $\mathcal{U}(\boldsymbol{\theta}_q)$ and $\mathcal{V}(\boldsymbol{\theta}_q)$ shown in the form $(1 - \text{RIC})$ for query parameter $\boldsymbol{\theta}_q = (0.10, 0.90)$. The red line corresponds to $(1 - \text{RIC}) = 10^{-7}$.

4.5. ROM Accuracy Analysis

The reduced-order model algorithm is applied with the following parameters and options:

- For global POD modes: $K_u^x = 40$, $K_u^c = 40$, $K_v^x = 50$, $K_v^c = 50$;
- For DA in (25), (30): local second order polynomial reconstruction, $M = 12$;
- For local POD modes: $m_u = 10$, $m_v = 10$;
- For DA in (45), (46): local first order polynomial reconstruction, $R = 2\Delta t$.

The resulting time-evolution of the three-dimensional capsule shape, that is reconstructed with the ROM model, is illustrated in Figure 12 for the query couple $\theta_q = (0.10, 0.90)$. The steady-state is reached before $\dot{\gamma}t = 3$, which explains that the capsule shape is the same for $\dot{\gamma}t = 3, 6, 9$.

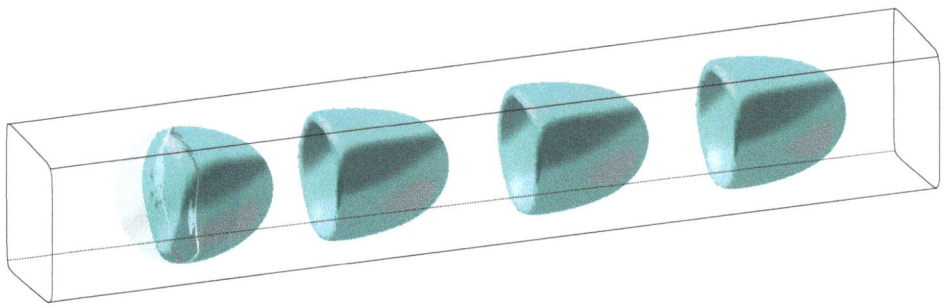

Figure 12. Three-dimensional shape of a capsule flowing in a square microchannel, reconstructed with the ROM model for $\theta = (Ca = 0.10, a/\ell = 0.90)$ and shown at $\dot{\gamma}t = 0, 0.4, 3, 6, 9$. The capsule initial shape is shown in transparency.

We now focus on the accuracy analysis of the proposed reduced-order model. The methodology for error measurement is based on a '*Leave-one-out*' cross-validation procedure, where each sample FOM solution is taken out from the database and then evaluated by the ROM model and compared to the original FOM one. The error is measured using the modified Hausdorff distance calculated on the capsule shapes at different instants.

Figure 13 shows the heat maps of the FOM-vs-ROM error computed over the parameter space at the time instants $\dot{\gamma}t = 1, 2, 4$ and 8. Figure 13 shows that the predicted ROM solutions are very accurate with a mean relative error below 0.2%. The maximum relative errors are below 3.5%: they occur along the boundary of the parameter domain, which is the only location where the predictions slightly lose in accuracy. This is probably due to a lack of well-distributed neighbors close to the boundaries, which affects the accuracy of the DA reconstruction (off-centre approximation). One can also notice that the accuracy of predictions decreases in time.

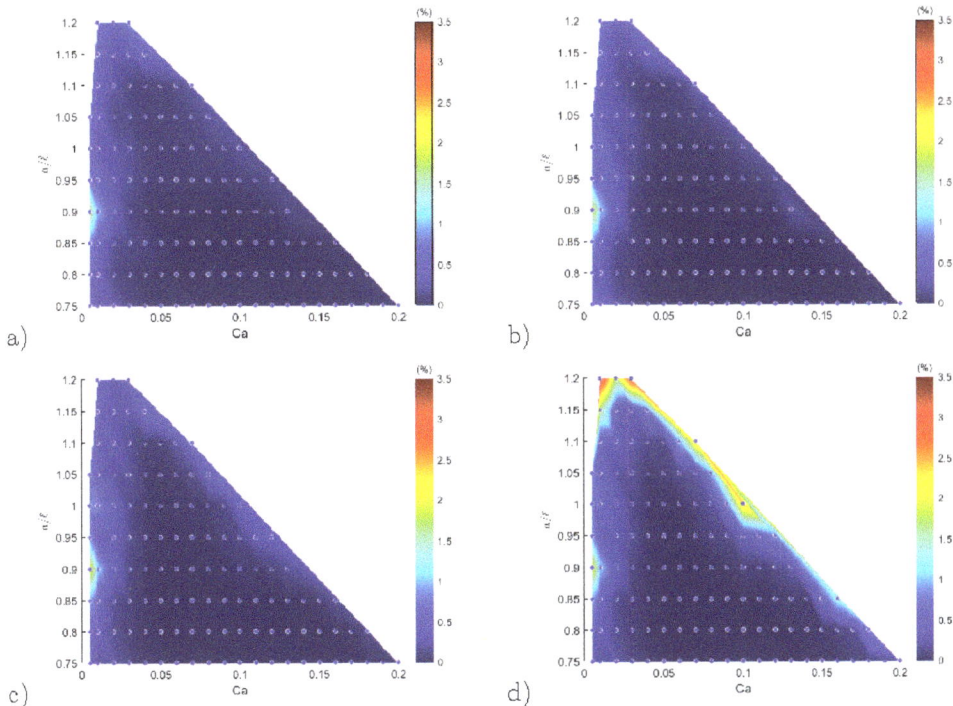

Figure 13. Heat maps of the normalized Hausdorff Distance d_{MH}/a of configuration prediction shapes over the parametric space at different transient states: (**a**) $\dot{\gamma}t = 1$; (**b**) $\dot{\gamma}t = 2$; (**c**) $\dot{\gamma}t = 4$; and (**d**) $\dot{\gamma}t = 8$. The maximum error is 3.26% in (**d**).

The capsule cross-section profiles predicted by the ROM (red dots) are compared to the reference FOM solutions (solid black line) in Figure 14 at different time instants ($\dot{\gamma}t = 0$, 1, 2 and 8) for the 6 configurations, selected as illustration in Figure 3. We observe that the reduced-order model returns very accurate solutions in terms of capsule shape as well-as axial position in the channel.

From the computing performance point of view, ROM-vs-FOM speedups are observed to be of order 10,000 with almost the same accuracy, making interactive exploration and real-time visual rendering possible.

4.6. CapsuleExplorer: Capsule Visualization/Exploration Software

We have developed an in-house software tool `CapsuleExplorer` based on the proposed ROM to provide the three-dimensional microcapsule deformation/evolution at any time $\dot{\gamma}t$ and for any θ_q in the admissible parameter domain. `CapsuleExplorer` allows one to select a particular couple $(Ca, a/\ell)$ in the admissible parameter domain, then to visualize the capsule dynamics between initial and final times, either in three dimensions or two dimensions with longitudinal or transversal cross-sectional view. The ROM high performance feature allows real-time exploration/visualization. `CapsuleExplorer` has been developed as a web application. Figures 15 and 16 show some screenshots of the graphics user interface, which will be useful for applications such as identifying the capsule wall mechanical properties through comparison with experimental results.

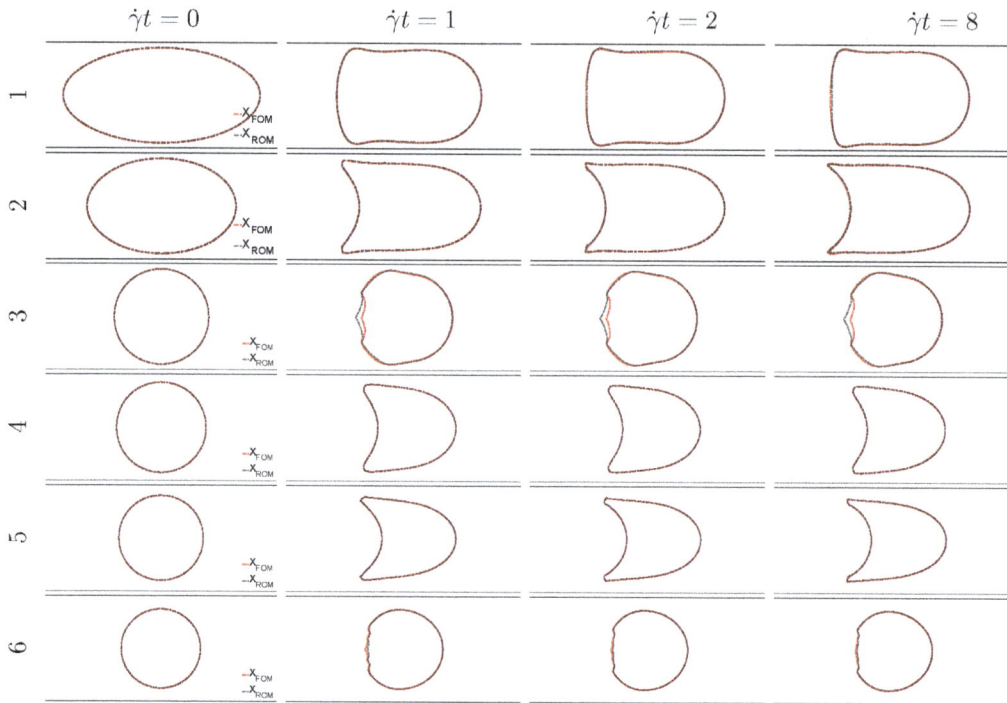

Figure 14. Comparison between the ROM (red dots) and FOM solutions (black line) of the capsule cross-section shapes in the plane $y = 0$ at the times $\dot{\gamma}t = 0, 1, 2$ and 8 respectively, for the 6 parameter couples selected in Figure 3. The horizontal lines correspond to the channel walls.

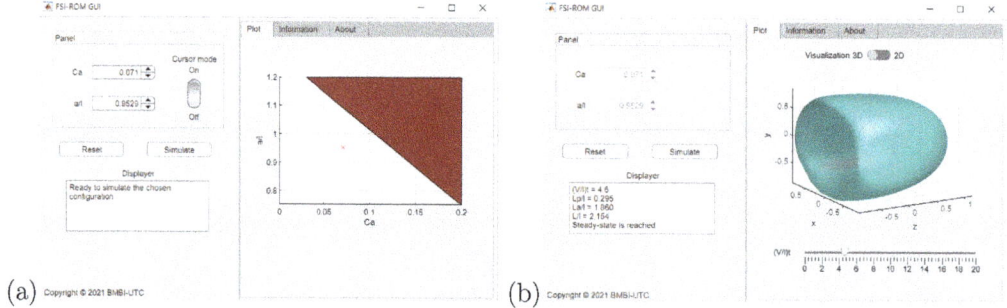

Figure 15. CapsuleExplorer: (**a**) parameter domain exploration; (**b**) dynamic 3D capsule view.

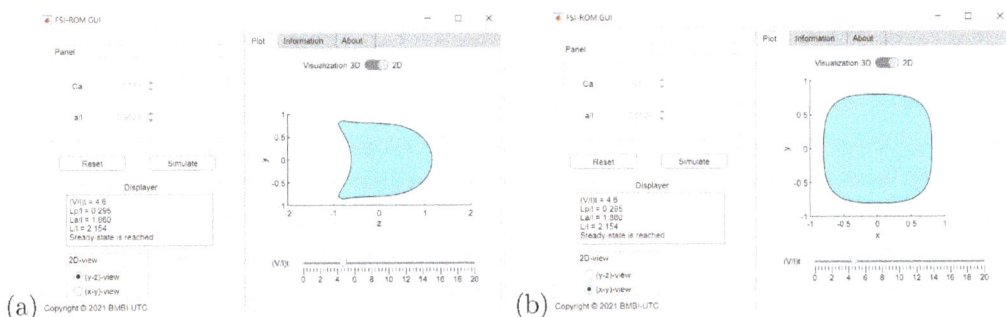

Figure 16. CapsuleExplorer: (**a**) dynamic 2D cross-section longitudinal view; (**b**) dynamic 2D cross-section transversal view.

5. Concluding Remarks

In this paper, we have presented an innovative data-driven reduced-order model that enables the dynamics of a deformable membrane flowing in a microchannel, from its initial state to the steady shape state. The ROM is built to be valid in a large domain of interest in the parameter plane $(Ca, a/\ell)$. Our FSI-ROM model first starts with an offline procedure to build two global orthonormal bases (space+parameter) that return good approximations of the FOM solutions over the whole parameter domain. The rather small truncation ranks already lead to an appreciable data dimensionality reduction, which is important for complexity and memory storage purposes.

The online stage consists in predicting the space-time solution for any query couple $\boldsymbol{\theta}_q = (Ca, a/\ell)$ in the parameter domain. In a first step, we determine a low-order basis for both the displacement and velocity vector variables. This is achieved by the use of diffuse approximation that returns an interpolated space-time solution at the query vector $\boldsymbol{\theta}_q$. Then an SVD analysis provides a suitable low-order spatial basis for final construction of the ROM in the second step. The physically-based ROM is made of the kinematics equation and the law of membrane quasi-static equilibrium in their reduced formulation. The unknown variables become the POD coefficient vectors of displacement and velocity fields. The reduced quasi-static equilibrium law is determined once again by the use of a diffuse approximation. The manifold learning is achieved by the use of time-snapshot data of the interpolated solution at $\boldsymbol{\theta} = \boldsymbol{\theta}_q$.

Numerical experiments confirm the efficiency of the method. ROM-vs-FOM speedups are observed to be of order 10,000 with almost the same accuracy (with less than a 0.3% error measured in terms of Hausdorff distance inside the parameter domain). Larger errors are encountered at the boundary of the parameter domain, but they still remain reasonable (up to 3.3% in Hausdorff distance). This work tends to show that model-order reduction techniques are complementary and valuable tools for the rapid design and optimization of capsules in healthcare engineering such as drug delivery through blood vessels.

The case of more complex FSI configurations such as the deformation of capsules flowing through a bifurcated microchannel will be investigated in a future work.

Author Contributions: Conceptualization, T.B., P.V., F.D.V. and A.-V.S.; Formal analysis, F.D.V., P.V.; Funding acquisition, A.-V.S.; Investigation, T.B., F.D.V. and A.-V.S.; Methodology, T.B., P.V. and F.D.V.; Project administration, A.-V.S.; Resources, C.Q.-G., C.D. and A.-V.S.; Software, T.B., C.Q.-G., C.D. and A.-V.S.; Supervision, F.D.V. and A.-V.S.; Validation, T.B.; Visualization, T.B.; Writing—original draft, T.B.; Writing—review & editing, C.Q.-G., C.D., P.V., F.D.V. and A.-V.S. All authors have read and agreed to the published version of the manuscript.

Funding: This project received funding from the European Research Council (ERC) under the European Union's Horizon 2020 research and innovation programme (Grant agreement No. ERC-2017-COG—MultiphysMicroCaps).

Conflicts of Interest: The authors declare no conflict of interest.

References

1. Williams, R.J.; Zipser, D. A learning algorithm for continually running fully Recurrent Neural Networks. *Neural Comput.* **1989**, *1*, 270–280. [CrossRef]
2. Raissi, M.; Perdikaris, P.; Karniadakis, G. Physics-informed neural networks: A deep learning framework for solving forward and inverse problems involving nonlinear partial differential equations. *J. Comput. Phys.* **2019**, *378*, 686–707. [CrossRef]
3. Peherstorfer, B.; Willcox, K. Dynamic data-driven reduced-order models. *Comput. Methods Appl. Mech. Eng.* **2015**, *291*, 21–41. [CrossRef]
4. Peherstorfer, B.; Willcox, K. Data-driven operator inference for nonintrusive projection-based model reduction. *Comput. Methods Appl. Mech. Eng.* **2016**, *306*, 196–215. [CrossRef]
5. Pawar, S.; Rahman, S.M.; Vaddireddy, H.; San, O.; Rasheed, A.; Vedula, P. A deep learning enabler for nonintrusive reduced order modeling of fluid flows. *Phys. Fluids* **2019**, *31*, 085101. [CrossRef]
6. Xiao, D.; Fang, F.; Pain, C.; Navon, I. A parameterized non-intrusive reduced order model and error analysis for general time-dependent nonlinear partial differential equations and its applications. *Comput. Methods Appl. Mech. Eng.* **2017**, *317*, 868–889. [CrossRef]
7. Cordier, L. Proper Orthogonal Decomposition: An overview. In *Lecture Series 2008-01 on Post-Processing of Experimental and Numerical Data*; Von Karman Institute for Fluid Dynamics: Sint-Genesius-Rode, Belgium, 2008
8. Benner, P.; Gugercin, S.; Willcox, K. A Survey of Projection-Based Model Reduction Methods for Parametric Dynamical Systems. *SIAM Rev.* **2015**, *57*, 483–531. [CrossRef]
9. Silva, D.F.; Alvaro, L.C. Practical implementation aspects of Galerkin reduced order models based on Proper Orthogonal Decomposition for Computational Fluid Dynamics. *J. Bras. Soc. Mech. Sci. Eng.* **2015**, *37*, 1309–1327. [CrossRef]
10. Gallivan, K.; Vandendorpe, A.; Van Dooren, P. Model Reduction of MIMO Systems via Tangential Interpolation. *SIAM J. Matrix Anal. Appl.* **2004**, *26*, 328–349. [CrossRef]
11. Chinesta, F.; Ladeveze, P.; Cueto, E. A short review on model order reduction based on Proper Generalized Decomposition. *Arch. Comput. Methods Eng.* **2011**, *18*, 395. [CrossRef]
12. Chinesta, F.; Ammar, A.; Cueto, E. On the Use of Proper Generalized Decompositions for Solving the Multidimensional Chemical Master Equation. *Rev. Européenne Mécanique Numérique/Eur. J. Comput. Mech.* **2010**, *19*, 53–64. [CrossRef]
13. Ghnatios, C.; Masson, F.; Huerta, A.; Leygue, A.; Cueto, E.; Chinesta, F. Proper Generalized Decomposition based dynamic data-driven control of thermal processes. *Comput. Methods Appl. Mech. Eng.* **2012**, *213–216*, 29–41. [CrossRef]
14. Barrault, M.; Maday, Y.; Nguyen, N.C.; Patera, A.T. An 'empirical interpolation' method: Application to efficient reduced-basis discretization of partial differential equations. *Comptes Rendus Math.* **2004**, *339*, 667–672. [CrossRef]
15. Chaturantabut, S.; Sorensen, D. Nonlinear model reduction via discrete empirical interpolation. *SIAM J. Sci. Comp.* **2010**, *32*, 2737–2764. [CrossRef]
16. Xiao, D.; Fang, F.; Buchan, A.; Pain, C.; Navon, O.; Du, J.; Hu, G. Non-linear model reduction for the Navier-Stokes equations using residual DEIM method. *J. Comp. Phys.* **2014**, *263*, 1–18. [CrossRef]
17. Lappano, E.; Naets, F.; Desmet, W.; Mundo, D.; Nijman, E. A greedy sampling approach for the projection basis construction in parametric model order reduction for structural dynamics models. In Proceedings of the 27th International Conference on Noise and Vibration Engineering, Leuven, Belgium, 19–21 September 2016; pp. 3563–3571.
18. Breitkopf, P.; Rassineux, A.; Villon, P. An Introduction to Moving Least Squares Meshfree Methods. *Rev. Européenne Eléments Finis* **2002**, *11*, 825–867. [CrossRef]
19. Amsallem, D.; Cortial, J.; Farhat, C. Toward real-time computational-fluid-dynamics-based aeroelastic computations using a database of reduced-order information. *AIAA J.* **2010**, *48*, 2029–2037. [CrossRef]
20. Audouze, C.; De Vuyst, F.; Nair, P.B. Reduced-order modeling of parameterized PDEs using time–space-parameter principal component analysis. *Int. J. Numer. Methods Eng.* **2009**, *80*, 1025–1057. [CrossRef]
21. Del Burgo, L.S.; Compte, M.; Aceves, M.; Hernández, R.M.; Sanz, L.; Álvarez-Vallina, L.; Pedraz, J.L. Microencapsulation of therapeutic bispecific antibodies producing cells: Immunotherapeutic organoids for cancer management. *J. Drug Target.* **2015**, *23*, 170–179.
22. Rabanel, J.M.; Banquy, X.; Zouaoui, H.; Mokhtar, M.; Hildgen, P. Progress technology in microencapsulation methods for cell therapy. *Biotechnol. Prog.* **2009**, *25*, 946–963. [CrossRef]
23. Chu, T.; Salsac, A.V.; Leclerc, E.; Barthès-Biesel, D.; Wurtz, H.; Edwards-Lévy, F. Comparison between measurements of elasticity and free amino group content of ovalbumin microcapsule membranes: Discrimination of the cross-linking degree. *J. Colloid Interface Sci.* **2011**, *355*, 81–88. [CrossRef]
24. Hu, X.Q.; Sévénié, B.; Salsac, A.V.; Leclerc, E.; Barthès-Biesel, D. Characterizing the membrane properties of capsules flowing in a square-section microfluidic channel: Effects of the membrane constitutive law. *Phys. Rev. E* **2013**, *87*, 063008. [CrossRef]
25. de Loubens, C.; Deschamps, J.; Georgelin, M.; Charrier, A.; Edwards-Lévy, F.; Leonetti, M. Mechanical characterization of cross-linked serum albumin microcapsules. *Soft Matter* **2014**, *10*, 4561–4568. [CrossRef] [PubMed]
26. Sévénié, B.; Salsac, A.V.; Barthès-Biesel, D. Characterization of Capsule Membrane Properties using a Microfluidic Photolithographied Channel: Consequences of Tube Non-squareness. *Procedia IUTAM* **2015**, *16*, 106–114. [CrossRef]

27. Gubspun, J.; Gires, P.Y.; Loubens, C.d.; Barthès-Biesel, D.; Deschamps, J.; Georgelin, M.; Leonetti, M.; Leclerc, E.; Edwards-Lévy, F.; Salsac, A.V. Characterization of the mechanical properties of cross-linked serum albumin microcapsules: Effect of size and protein concentration. *Colloid Polym. Sci.* **2016**, *294*, 1381–1389. [CrossRef]
28. De Loubens, C.; Deschamps, J.; Edwards-Lévy, F.; Leonetti, M.; Raissi, M.; Perdikaris, P.; Karniadakis, G.E. Physics-informed neural networks: A deep learning framework for solving forward and inverse problems involving nonlinear partial differential equations. *J. Fluid Mech.* **2016**, *378*, 686–707.
29. Quesada, C.; Dupont, C.; Villon, P.; Salsac, A.V. Diffuse approximation for identification of the mechanical properties of microcapsules. *Math. Mech. Solids* **2021**, *26*, 1018–1028. [CrossRef]
30. Wang, Z.; Sui, Y.; Salsac, A.V.; Barthès-Biesel, D.; Wang, W. Motion of a spherical capsule in branched tube flow with finite inertia. *J. Fluid Mech.* **2016**, *806*, 603–626. [CrossRef]
31. Vesperini, D.; Chaput, O.; Munier, N.; Maire, P.; Edwards-Lévy, F.; Salsac, A.V.; Le Goff, A. Deformability- and size-based microcapsule sorting. *Med. Eng. Phys.* **2017**, *48*, 68–74. [CrossRef] [PubMed]
32. Wang, Z.; Sui, Y.; Salsac, A.V.; Barthès-Biesel, D.; Wang, W. Path selection of a spherical capsule in a microfluidic branched channel: Towards the design of an enrichment device. *J. Fluid Mech.* **2018**, *849*, 136–162. [CrossRef]
33. Kabacaoğlu, G.; Biros, G. Sorting same-size red blood cells in deep deterministic lateral displacement devices. *J. Fluid Mech.* **2019**, *859*, 433–475. [CrossRef]
34. Häner, E.; Vesperini, D.; Salsac, A.V.; Le Goff, A.; Juel, A. Sorting of capsules according to their stiffness: From principle to application. *Soft Matter* **2021**, *17*, 3722–3732. [CrossRef]
35. Quesada, C.; Villon, P.; Salsac, A.V. Real-time prediction of the deformation of microcapsules using Proper Orthogonal Decomposition. *J. Fluids Struct.* **2021**, *101*, 103193. [CrossRef]
36. Savignat, J.M. Approximation Diffuse Hermite et Ses Applications. Ph.D. Thesis, École Nationale Supérieure des Mines de Paris, Paris, France, 2000.
37. Raghavan, B.; Breitkopf, P.; Tourbier, Y.; Villon, P. Towards a space reduction approach for efficient structural shape optimization. *Struct. Multidiscip. Optim.* **2013**, *48*, 987–1000. [CrossRef]
38. Hu, X.Q.; Salsac, A.V.; Barthès-Biesel, D. Flow of a spherical capsule in a pore with circular or square cross-section. *J. Fluid Mech.* **2012**, *705*, 176–194. [CrossRef]
39. Wang, X.; Merlo, A.; Dupont, D.; Salsac, A.V.; Barthès-Biesel, D. Characterization of the mechanical properties of microcapsules with a reticulated membrane: Comparison of microfluidic and microrheometric approaches. *Flow* **2021**, in press.
40. Walter, J.; Salsac, A.V.; Barthès-Biesel, D.; Le Tallec, P. Coupling of finite element and boundary integral methods for a capsule in a Stokes flow: Numerical Methods for A Capsule in A Stokes Flow. *Int. J. Numer. Methods Eng.* **2010**, *83*, 829–850. [CrossRef]
41. Barthès-Biesel, D. Modeling the motion of capsules in flow. *Curr. Opin. Colloid Interface Sci.* **2011**, *16*, 3–12. [CrossRef]
42. Golub, G.H.; Reinsch, C. Singular value decomposition and least squares solutions. In *Linear Algebra*; Bauer, F.L., Ed.; Springer: Berlin/Heidelberg, Germany, 1971; pp. 134–151.
43. Buhmann, M.D. *Radial Basis Functions*; Cambridge Monographs on Applied and Computational Mathematics; Cambridge University Press: Cambridge, UK, 2003.
44. Dubuisson, M.P.; Jain, A. A modified Hausdorff distance for object matching. In Proceedings of the 12th International Conference on Pattern Recognition, Jerusalem, Israel, 9–13 October 1994; Volume 1, pp. 566–568.

Article

Study of Concentrated Short Fiber Suspensions in Flows, Using Topological Data Analysis

Rabih Mezher [1], Jack Arayro [1], Nicolas Hascoet [2] and Francisco Chinesta [2,*]

[1] College of Engineering and Technology, American University of the Middle East, Egaila 54200, Kuwait; rabih.mezher@aum.edu.kw (R.M.); Jack.Arayro@aum.edu.kw (J.A.)
[2] PIMM Lab, ESI Group Chair, Arts et Metiers Institute of Technology, 151 Boulevard de Hopital, 75013 Paris, France; Nicolas.Hascoet@ensam.eu
* Correspondence: Francisco.Chinesta@ensam.eu

Abstract: The present study addresses the discrete simulation of the flow of concentrated suspensions encountered in the forming processes involving reinforced polymers, and more particularly the statistical characterization and description of the effects of the intense fiber interaction, occurring during the development of the flow induced orientation, on the fibers' geometrical center trajectory. The number of interactions as well as the interaction intensity will depend on the fiber volume fraction and the applied shear, which should affect the stochastic trajectory. Topological data analysis (TDA) will be applied on the geometrical center trajectories of the simulated fiber to prove that a characteristic pattern can be extracted depending on the flow conditions (concentration and shear rate). This work proves that TDA allows capturing and extracting from the so-called persistence image, a pattern that characterizes the dependence of the fiber trajectory on the flow kinematics and the suspension concentration. Such a pattern could be used for classification and modeling purposes, in rheology or during processing monitoring.

Keywords: topological data analysis (TDA); reinforced polymers; concentrated suspensions; flow induced orientation; discrete numerical simulation

1. Introduction

Reinforced polymers are widely used in industry for enhancing mechanical and functional performances while keeping the cost reasonable. The main issue related to the use of fiber-based reinforced polymers for elaborating short fiber composites is due to the difficulty of accurately predicting the flow induced orientation, with the final properties becoming strongly dependent on the final orientation state of fibers in the formed part.

The orientation evolution of an ellipsoidal fiber immersed in a flow characterized by a gradient of velocity can be computed by using the so-called Jeffery equation [1]. However, as soon as the fiber concentration increases, intense interactions between the rotating fibers takes place and the orientation kinematics of each fiber will differ from the one predicted by the Jeffery model.

At the population level (ensemble of fibers in a representative volume in which the velocity gradient is assumed almost identical) the interactions can be described as a diffusion term acting on the fiber orientation probability distribution Ψ, whose evolution is governed by the so-called Fokker-Planck equation [2], and more concretely the so-called Folgar-Tucker model [3]. Due to the fact that the orientation distribution depends on the physical coordinates (space and time) and also on the configurational ones (the orientation $\mathbf{p}$ defined on the surface of the unit sphere), $\Psi(x,t,\mathbf{p})$, descriptions based on the moments of the orientation distribution are preferred [2,4]. Thus, the second order moment of the orientation distribution function reads:

$$\mathbf{a}(x,t) = \oint \mathbf{p} \otimes \mathbf{p} \Psi(x,t,\mathbf{p})\, d\mathbf{p}, \qquad (1)$$

where $\otimes$ refers to the tensor product.

When considering a description based on the orientation tensors (orientation distribution moments), the diffusion term describing fiber interaction within the Folgar & Tucker formulation, results in a sort of randomizing term that tends, to evolve the orientation towards the isotropic state, that is $\mathbf{a} \to \mathbf{I}/3$ (in 3D), with $\mathbf{I}$ the identity tensor [2,4].

However, many hypotheses were introduced when deriving the models describing the fiber interaction, fact that limits their validity and accuracy. Discrete simulations consider a population of fibers, subjected to two main actions, the hydrodynamic ones induced by the fluid flow, flow that is assumed unperturbed by the fibers presence and their orientation state, and the forces that apply when two neighbor fibers approach mutually activating, first hydrodynamics forces and then contact forces for avoiding interpenetration.

Discrete simulations are extremely expensive because of the high number of fibers to be considered for representing the different concentration regimes, and because of the extremely small time steps that the small length scales involved by the fibers interaction imply.

When fibers enter in contact, having a non-null relative velocity, the interaction will affect the orientation kinematics from one side, but it will also affect the fibers geometrical center trajectory. Thus, it is postulated that this trajectory will depend on the number and intensity of the fiber interaction, both expected scaling with the flow gradient of velocity, the fiber concentration and the orientation state.

Thus, the analysis of those erratic trajectories that the fiber follow, should provide a very valuable information on the orientation state (difficult to measure in 3D flows of concentrated suspensions), the local concentration that could differ from one point to another in the flow, or even the effective velocity gradient that could differ from the nominal one, that as previously indicated is assumed the one unperturbed by the fibers presence.

However, extracting information from those erratic trajectories seems difficult, needing the use of adequate metrics to compare them, that apparently seem very different even when the flow conditions remain identical. Moreover, the usual statistical descriptors (widely considered for describing roughness for instance) seem insufficient for describing the trajectory richness. Thus, robust metrics for describing in a concise, compact and rich enough way, with the suitable invariance properties, are needed for making possible unsupervised clustering and supervised classification of the different trajectories. For that purpose, the present work considers topological data analysis for analyzing the stochastic time-series induced by the fiber interactions, at the level of the movement of the fiber geometrical center.

The paper is organized as follows. Section 2 describes the discrete simulation of flows involving concentrated fiber suspensions. Then, in Section 3, the so-called Topological Data Analysis (TDA) will be revisited. Finally, the numerical results will be reported in Section 4, before addressing some final concluding remarks in Section 5.

2. Discrete Simulation

The main assumptions considered in the the modelling and simulation framework are [5,6]:

1. The suspending fluid is Newtonian, incompressible and the flow is laminar;
2. The fluid velocity gradient is assumed being homogeneous in the considered representative volume where the calculations are performed, with the velocity field assumed unperturbed by the particles presence and their orientation;
3. The mass of the fibers is negligible, thus the inertia of the fibers is neglected;
4. Fibers are considered to have the same length, but they could have different length;
5. The long-range hydrodynamic interactions are considered along with short-range hydrodynamic interactions between fibers;
6. Initially and before the simulation starts, the fibers are homogeneously and almost isotropically distributed in the considered volume, with interpenetration prevented.

The position of the geometrical center G of fiber (α), $\mathbf{r}^{(\alpha)}$, is given by

$$\mathbf{r}^{(\alpha)} = x^{(\alpha)}\mathbf{x} + y^{(\alpha)}\mathbf{y} + z^{(\alpha)}\mathbf{z}, \tag{2}$$

where $\mathbf{x}$, $\mathbf{y}$ and $\mathbf{z}$ represent respectively the three unit vectors related to the three space coordinates.

Fibers are assumed having an ellipsoidal shape, with length l and diameter d (taken at the axis center). Thus, the aspect ratio of the fibers r reads:

$$r = \frac{l}{d}. \tag{3}$$

In the numerical simulations described later, the considered fibers have an aspect ratio of 20. Thus, the fibers will be represented by elongated ellipsoids, whose orientation will be described by a unit vector $\mathbf{p}$ aligned with the ellipsoid longest axis. Moreover, the considered aspect ratio allows assuming the fibers rigid, as experimental observations prove for usual materials, as for example glass fibers.

Since the suspensions are considered concentrated, with the fiber volume fraction noted by ϕ, the following inequality applies:

$$\phi \geq \frac{1}{r}. \tag{4}$$

The higher r (i.e., long fibers), the more the system is considered concentrated for a fixed fibers concentration ϕ. In what follows, the fibers are supposed to be sufficiently long (i.e., $r \gg 1$), approaching the cylindrical shape.

The fixed frame is defined from $(O, \mathbf{x}, \mathbf{y}, \mathbf{z})$, whereas another frame is attached to each fiber: $(G, \mathbf{x}', \mathbf{y}', \mathbf{z}')$. A shear flow is applied, with the velocity field expressed from

$$\mathbf{V}^T(\mathbf{x}) = (V_1, V_2, V_3) = (\dot{\gamma}y, 0, 0), \tag{5}$$

with $\dot{\gamma}$ the applied shear rate and y the y-coordinate of the fiber geometrical center. This expression allows defining the velocity gradient $\nabla \mathbf{V}$ as well as its symmetric and skew-symmetric parts, $\mathbf{D}$ and $\mathbf{W}$ respectively, with $\mathbf{\Omega} = \frac{1}{2}(\nabla \times \mathbf{V})$.

The fiber orientation is defined by the unit vector $\mathbf{p}^{(\alpha)}$ such as $\mathbf{p}^{(\alpha)} = p_1^{(\alpha)}\mathbf{x} + p_2^{(\alpha)}\mathbf{y} + p_3^{(\alpha)}\mathbf{z}$. The relative fluid/particle velocity at G reads

$$\dot{\mathbf{q}}^{(\alpha)} = \dot{\mathbf{r}}^{(\alpha)} - \mathbf{V}(\mathbf{r}^{(\alpha)}) = \dot{\mathbf{r}}^{(\alpha)} - \dot{\gamma}y^{(\alpha)}\mathbf{x}. \tag{6}$$

2.1. Fiber Motion Equations: Translation

The net force that the fluid transfer to the fiber scales with the relative velocity at G from the so-called resistance tensor $\boldsymbol{\zeta}$, and then the force balance with the acting force $\mathbf{F}$, reads

$$\mathbf{F}^{(\alpha)} + \boldsymbol{\zeta}^{(\alpha)} \cdot \dot{\mathbf{q}}^{(\alpha)} = 0, \tag{7}$$

where the friction tensor expression is given in [7], and depends on the fluid viscosity, the fiber geometry and its orientation.

2.2. Fiber Motion Equations: Rotation

First we consider the dilute case where fiber interaction cans be neglected. The fluid deformation induces on the fiber the torque $\mathcal{H}^{(\alpha)} : \mathbf{D}$ (with $\mathcal{H}$ a third order resistance tensor) and the fluid/fiber relative rotary velocity $\boldsymbol{\omega}^{(\alpha)}$ induces the torque $\boldsymbol{\xi}^{(\alpha)} \cdot \boldsymbol{\omega}^{(\alpha)}$, with $\boldsymbol{\xi}^{(\alpha)}$ a second order resistance tensor. Both resistance tensors [7] depend again on the fluid viscosity, fiber geometry and fiber orientation.

When neglecting inertia effects, the torque balance (in absence of fiber interactions) reads

$$\boldsymbol{\zeta}^{(\alpha)} \cdot \boldsymbol{\omega}^{(\alpha)} + \boldsymbol{\mathcal{H}}^{(\alpha)} : \mathbf{D} = 0, \qquad (8)$$

from which the fiber rotary velocity can be extracted,

$$\dot{\mathbf{p}}^{(\alpha)} = -\mathbf{p}^{(\alpha)} \times \left(\boldsymbol{\omega}^{(\alpha)} - \boldsymbol{\Omega} \right), \qquad (9)$$

that for infinite aspect ratio fibers leads to

$$\dot{\mathbf{p}}^{(\alpha)} = \dot{\mathbf{p}}_J^{(\alpha)} = \mathbf{W} \cdot \mathbf{p}^{(\alpha)} + \left[\mathbf{D} \cdot \mathbf{p}^{(\alpha)} - \left(\mathbf{D} : \mathbf{p}^{(\alpha)} \otimes \mathbf{p}^{(\alpha)} \right) \mathbf{p}^{(\alpha)} \right], \qquad (10)$$

that coincided with the Jeffery equation [1].

When the suspension becomes concentrated enough, fiber-fiber interactions occur. Thus, short-range forces will appear on the fibers as they interact.

There are two types of interactions considered via two types of forces: A lubrication force F_{lb} occurs when two fibers approach one another; and a contact force F_c when they touch, that when neglecting friction (the roughness of the fiber surface is very small, fact that enables neglecting the induced friction force), the contact force, as well as the lubrication one, is assumed acting in the normal direction.

The resulting interaction force on fiber (α) reads:

$$\mathbf{F}^{(\alpha)} = \sum_{\beta \neq \alpha} F_c^{(\alpha,\beta)} \mathbf{n}^{(\alpha,\beta)} + \sum_{\mu \neq \alpha} F_{lb}^{(\alpha,\mu)} \mathbf{n}^{(\alpha,\mu)}, \qquad (11)$$

that will induce a torque $\mathbf{T}^{(\alpha)}$ on the considered fiber, leading to the torque balance

$$\mathbf{T}^{(\alpha)} + \boldsymbol{\zeta}^{(\alpha)} \cdot \boldsymbol{\omega}^{(\alpha)} + \boldsymbol{\mathcal{H}}^{(\alpha)} : \mathbf{D} = 0, \qquad (12)$$

from which

$$\boldsymbol{\omega}^{(\alpha)} = -\boldsymbol{\zeta}^{(\alpha)-1} \cdot \left(\mathbf{T}^{(\alpha)} + \boldsymbol{\mathcal{H}}^{(\alpha)} : \mathbf{D} \right), \qquad (13)$$

leading to the fiber rotary velocity $\dot{\mathbf{p}}^{(\alpha)}$.

Thus, knowing the resulting force applied on fiber (α) one can compute the relative velocity at G, $\dot{\mathbf{q}}^{(\alpha)}$ (that allows updating the fiber center position), and the fiber rotary velocity $\dot{\mathbf{p}}^{(\alpha)}$.

The calculation of the distance between two fibers and the calculation of the lubrication forces depending on the approaching velocity $\dot{\Theta}^{(\alpha,\beta)}$, were detailed in [8].

Contact forces are assumed to occur if the gap between two close fibers is equal to zero and if $F_c^{(\alpha,\beta)} \neq 0$. The condition employed in the present work reads [9]

$$\frac{d}{dt} \left[\left(\mathbf{r}^{(\alpha)} - \mathbf{r}^{(\beta)} \right) \cdot \mathbf{n}^{(\alpha,\beta)} \right] = \dot{\Theta}^{(\alpha,\beta)} = 0, \qquad (14)$$

with $\dot{\Theta}^{(\alpha,\beta)} \approx 0$. It physically means that two fibers in contact cannot penetrate one another.

For solving the problem, fibers are grouped. Imagine that fibers (α) and (β) are in contact. The first group is composed by all the fibers in interaction with fiber (α). The second group is all the fibers in interaction with fiber (β). There is one unknown force for each pair of fibers, because the forces acting on the two fibers are equal in magnitude but opposite in direction. All forces for these two groups are coupled and should be solved together with all the interactions in the suspension by enforcing the kinematic constraints (14) at each contact level. For additional details the interested reader can refer to [5] and the references therein.

3. Topological Data Analysis

Data is generated by considering a population of fibers inside a computational box, that represents the so-called representative volume. The number of fibers depends on the considered fibers volume fraction (fibers concentration). Then, a simple shear flow is assumed taking place inside, with the velocity given by $\mathbf{V}^T = (\dot{\gamma}y, 0, 0)$. As discussed above, the flow is assumed unperturbed by the fibers presence and orientation. In absence of interactions, the geometrical center of each fiber will follow a rectilinear trajectory, traversing the computational box, until leaving it from its right boundary. Instead of increasing the box size, fully periodic boundary conditions are enforced. Thus, as soon as a fiber leaves the box from its right boundary it is re-injected into the box through its left boundary. The computational cell perfectly represents the bulk flow conditions, as soon as the analyzed flow is not affected by the physical walls (e.g., the mould walls). Here, we assume that the flow cell (representative volume) is far enough (with respect to the fiber length) from the physical walls for ignoring the effects of those walls.

In the absence of interactions, the orientation of each fiber describes the so-called Jeffery orbit. When concentration increases, Jeffery orbits intersect one another and then lubrication and contact forces appear when fibers interact. The number of interactions will scale with the fiber concentration, while the interaction intensity scales with the applied shear rate. Thus, the higher the fibers' concentration and the applied shear rate, the more intense and frequent the interactions occurring in the flow, creating a strong perturbation in the orientation kinematics (fiber rotary velocity) as well as in the erratic trajectory described by the fiber centers.

The interactions (lubrication and contacts) occur inside the box, but due to the assumed and enforced periodic boundary conditions, fibers located in the neighborhood of the right boundary can interact with the ones located in the neighborhood of the left one, and those close to the bottom boundary with the ones close to the top one, and similarly for the front and rear sides of the box.

Fibers are initially located randomly into the box, while avoiding interpenetration. Thus, at the end of the box filling an almost isotropic orientation state is obtained, i.e., $\mathbf{a} \approx \mathbf{I}/3$.

A test fiber is considered close to the center of the box, and its trajectory is recorded, in particular the three components of the fluctuating vector $\dot{\mathbf{q}}$ acting on it, that will represent the three time series S_x, S_y, S_z: $S_x = \{\dot{q}_x^1, \dot{q}_x^2, ...\}$ and similarly for the other two times series. In these time series and for comparison purposes, the exponent refers to the quantity of applied strain, i.e., $\bullet^n$ refers to $\bullet(\dot{\gamma}t_n)$.

The kinematics of the test particle is followed a certain time, in order to almost cover the three main regimes that it is experiencing:

1. The first regime is the one taking place at the very beginning when the flow starts, where the initial fiber distribution evolves in absence of interactions, until fibers approaching ones another induce the expected fiber interaction (lubrication and contact);
2. The second regime is the one when the orientation of the fibers in the population evolve, trying to align with the flow direction (induced by the applied shear) but in presence of numerous and intense interactions;
3. The third region is an almost stablised regime, when fibers are quite aligned with the preferential orientation direction (the x-coordinate in the case here studied). In this case the number of interaction reduces because when fibers are almost aligned in the same direction, interactions are much less probable and much less intense. However, as the fibers are ellipsoids, they cannot align in a stable manner in the flow direction, the rotary velocity never vanishes, even in absence of interactions. The rotary velocity becomes very small when ellipsoids align along the flow direction, and consequently the fiber spend a lot of time aligned in the flow direction, but it continues its rotation, and the rotary velocity increases when the orientation moves apart form the flow direction, reaching its maximum velocity when the y-coordinate reaches its maximum value. Then, the rotary velocity decreases again when the fiber orientation approaches again the flow direction, and the cycle repeats and rotation continues. Thus, the fiber

spend long periods almost aligned with the flow, and rotate very fast outside this most stable direction (the local orientation with the flow). During this fast rotation the interactions are numerous and intense, because of the fact that each fiber rotates at different instants.

In order to compare the just referred time series, we must consider appropriate metrics able to find the similarity of times series, neither identical nor superposable. Topological Data Analysis [10–12] inherits the invariance properties of topology, and then it is an appealing candidate for analyzing, describing and finally classifying time series with respect to the concentration regime and the applied shear rate.

For the sake of clarity, we will consider a generic time series $S = \{s^1, s^2, ..., s^k, ...\}$. To extract the topology of the data composing the time series, first the extremum points (local minimums and local maximums) are identified, and then we proceed to the one-to-one local-minimum/local-maxixum neighbors pairing. In the pairing process, when multiple alternatives exist, the one maximizing the max to min distance is retained.

Now, we assume that P min-max pairs have been constructed: $(b^1, d^1), ..., (b^P, d^P)$, where b refers to the minimum, also referred as *birth*, and d refers to the maximum, or the *death*. Each one of these pairs results in a point in the so-called persistence diagram, with the birth component reported in its horizontal axis and the deaths in the vertical one. Being the maximum always greater (or equal) than the minimum, points will group in the upper part with respect to the bisector (diagonal of the square birth/death representation).

Instead of representing on the vertical axis the deaths, an alternative derived representation consists of representing the lifetime, that is, $d^k - b^k$. Thus, the points reported into the so-called life-time diagram are the P data points: $(b^1, d^1 - b^1), ..., (b^P, d^P - b^P)$, that now appear distributed everywhere in the 2D representation.

Calculating distances between clouds of points is possible when using an adequate metrics. One possibility consists of using the Wasserstein metrics usually employed in optimal transport [13], that first matches the points of all the considered sets, in order to minimize the cost related to the distance among them, and then compute the Euclidian distance between the matched points-sets.

However, using this kind of data representation in usual artificial intelligence and machine learning techniques for clustering, classifying and modeling (i.e., constructing regressions) remains its trickiest issue. For that reason, a step forward consists of transforming the life-time diagram into the so-called persistence image, defined in a vector space facilitating its post-processing for a diversity of purposes.

For that purpose, and as described in our former works [14–16], we associate to each data-point in the life-time diagram a bivariate normal distribution, weighted and then integrated in different patches on a square domain covering the support of the regularized life-time diagram, leading to the so-called persistence images.

The resulting persistence images have an important property, the one of be invariant for time-series having similar topologies, even when they cannot be perfectly matched when using their time-representations.

Thus, persistence images enable efficient unsupervised clustering or supervised classification, and can be used also as input in regressions, by considering convolutional neural networks –cNN– directly applied on them, or nonlinear polynomial regression applied on the coefficients of their PCA decomposition [17].

4. Results

According with the rationale described at the beginning of Section 3, different time series related to the movement of a test fiber in different flow conditions, the last characterized by the fibers volume fraction (%) and the applied shear rate s^{-1}, were generated. The considered design of experiments –DoE– is given in Table 1.

The initial orientation is almost isotropic, that is, there are fibers pointing in any direction of the unit sphere, with an almost a uniform distribution. Thus, one expects that when the flow starts, the flow induced orientation, trying to align all the fibers along the flow direction (as discussed before), will create frequent and intense fiber-fiber interactions, scaling with the shear rate and the concentration. These interactions will induce significant displacements of the fibers geometrical centers. When the fibers align along the flow direction, i.e., with the velocity field, they remain most of the time aligned with the flow. However the alignment with the flow is never permanent because of two main reasons.

Table 1. Design of Experiments.

Case	Concentration (%)	Shear Rate (s^{-1})
1	14	5
2	18	5
3	22	5
4	18	1
5	18	3
6	18	7

First, when considering fibers modeled by ellipsoids (as it is the case here) the local alignment is not a steady solution (no steady solution exists). The rotary velocity reaches its smallest value when the fiber is aligned with the flow, but it is not exactly zero. Thus, the fiber moves apart from the alignment with the flow, to make a turn, coming back to the alignment with the flow, where again it spends a long period before starting another rotation, and so on.

The second advocated reason is, that even if the interaction is much less intense when fibers are globally quite aligned with the flow, the sporadic rotations just described create fiber-fiber interactions that induce the displacement of the fibers geometrical centers, while the orientation also deviate from the local alignment with the flow.

Moreover, as fibers are rotating according to the applied shear, in the clockwise direction in our case, the displacement of the fibers geometrical center is expected exhibiting an asymmetric behavior.

For confirming the previous expectations, we consider Cases 4 and 6 in Table 1, related to the minimum and maximum applied shear rates, both having the same fiber concentration, and compares the displacement of the test fiber geometrical center along the y-direction (the shear direction), both cases represented respectively in Figures 1 and 2. These two figures prove that the larger is the shear rate, the higher is the fiber-fiber interaction intensity, and consequently the displacement induced on the fibers along the shear direction (y-direction—with the flow occurring along the x-direction).

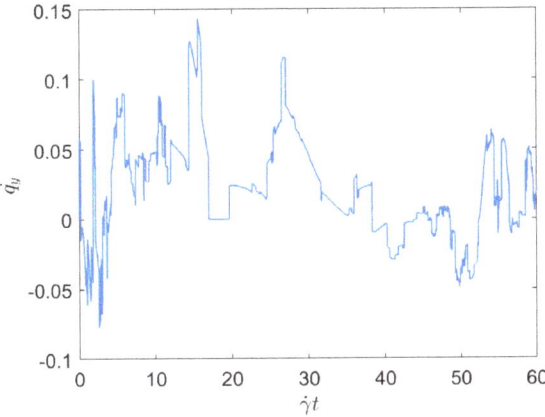

Figure 1. Time series related to displacement on the y-direction in Case 4: Minimum shear rate.

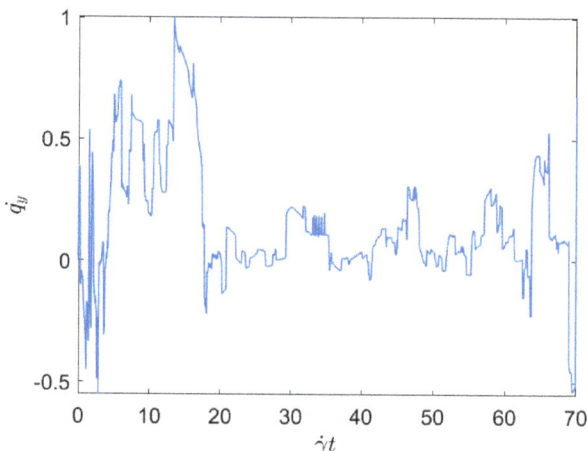

Figure 2. Time series related to displacement on the y-direction in Case 6: Maximum shear rate.

To evaluate the effect of the concentration while keeping constant the applied shear, we consider Case 1 and Case 3, with respectively the minimum and maximum fibers concentration (both subjected to the same applied shear rate). Figures 3 and 4 represent the associated displacement along the shear direction (y-coordinate). As it can be noticed from the observation of these figures, for the lower concentration, after the numerous interactions that follow the flow initiation, a plateau corresponding to the fibers alignment along the flow direction, where interactions almost disappear, is noticed. As discussed previously, fibers move apart form the local alignment for performing a full rotation before coming back again to the orientation with the flow, in which it stays for a long period (the rotary velocity is minimum when fibers are almost aligned with the flow). For the maximum concentration, fiber-fiber interactions persist after the transient regime, and the permanent regime continues exhibiting intense fluctuations induced by the interactions. It can be stressed that the concentration mainly affects the number of interactions, but their intensity seems more influenced by the shear rate than by the fiber concentration.

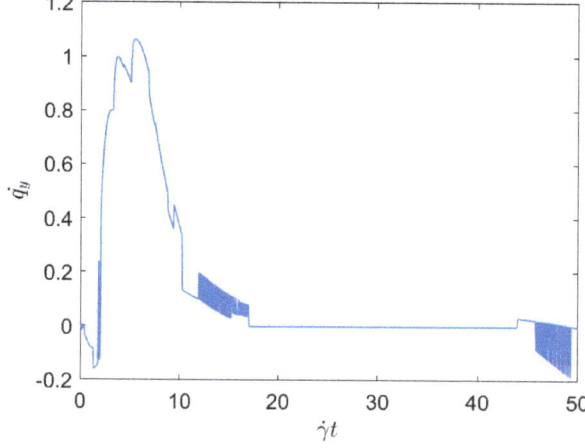

Figure 3. Time series related to displacement on the y-direction in Case 1: Minimum concentration.

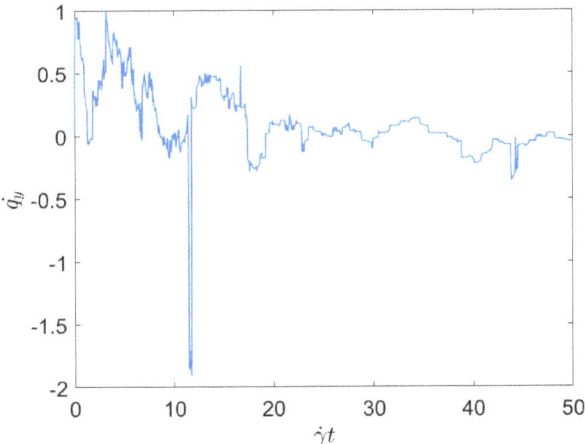

Figure 4. Time series related to displacement on the y-direction in Case 3: Maximum concentration.

To better appreciate the number and distribution of the topological events, Figures 5–8 show the persistence diagrams related respectively to Figures 1–4, where each blue dot represents a topological event, with its appearance reported in the x-coordinate axis and its death in its y-coordinate axis, being the vertical distance to the diagonal a representation of its persistence (its lifetime).

Figures 5 and 6 clearly reveal that the topology becomes more persistent when increasing the shear rate, with the associated topological event appearance asymmetrically distributed with respect to the zero value. High shear rates induce strong interactions (as observed in Figure 2) that result in highly fluctuating dynamics, with large amplitudes, that result in persistent topology. On the contrary, when the shear rate decreases the fluctuations are much less intense (smaller amplitudes) inducing an ephemeral topology, with the topological events closer to the diagram diagonal.

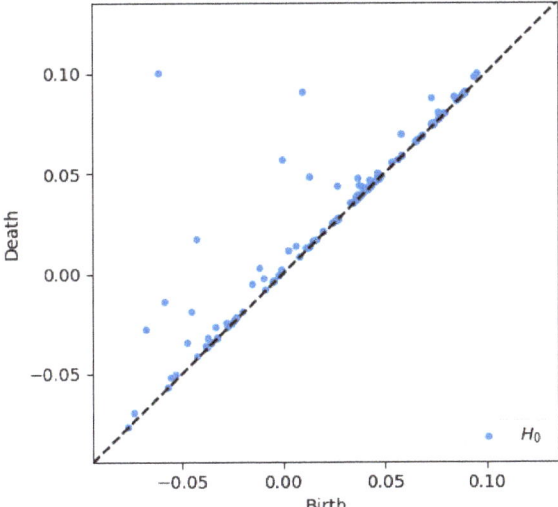

Figure 5. Persistence diagram related to displacement on the y-direction in Case 4: Minimum shear rate.

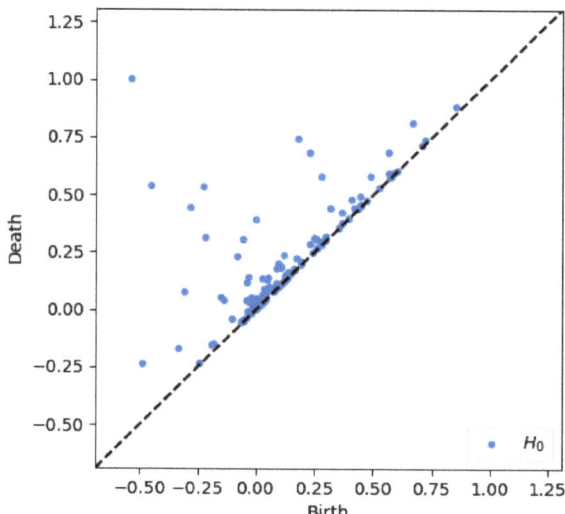

Figure 6. Persistence diagram related to displacement on the y-direction in Case 6: Maximum shear rate.

Now, focusing on the effect of concentration, from Figures 7 and 8 it can be stressed that in the dilute regime, represented by Figure 3, the largest persistent topology is associated with the transient regime, with ephemeral events occurring as soon as fibers almost align with the flow.

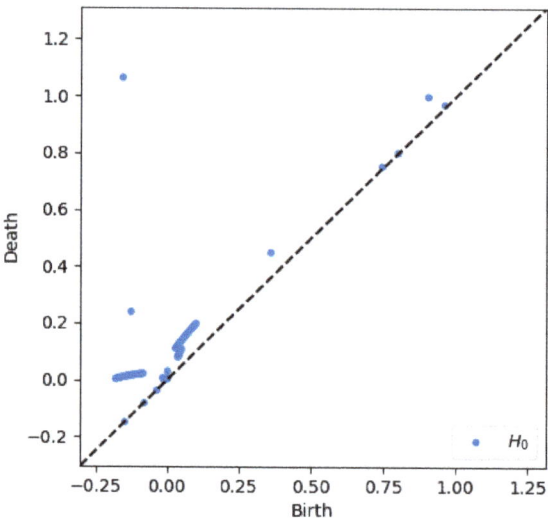

Figure 7. Persistence diagram related to displacement on the y-direction in Case 1: Minimum concentration.

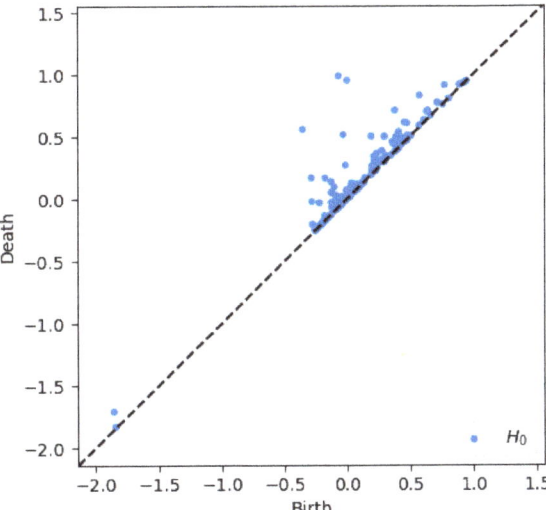

Figure 8. Persistence diagram related to displacement on the y-direction in Case 3: Maximum concentration.

When the concentration increases nothing changes significantly, as expected, concerning the most persistent topology, however, the ephemeral one becomes more abundant and erratic than the one related to the dilute case. It is important to notice that the scale of representation is impacted by an isolated negative displacement that induces a displacement along the x-axis, and that could be considered as an outlier. These findings confirm that the concentration affects more the ephemeral events that the persistent topology.

Thus, two main scales can be differentiate, the one related to the transient regime, involving more persistent topology, and the one related to long-time regime exhibiting more ephemeral events.

The main issue, previously discussed, is the way of using a compact, concise and complete descriptor of the time series depicted in the previous figures (Figures 1–4), more easy to manipulate than the discrete persistence diagrams reported in Figures 5–8.

The use of persistence images is a valuable route for accomplishing it, because they allow extracting and differentiating micro and macro events, inducing ephemeral or persistent topology. Persistence images are defined in a vector space and can be easily manipulated by most of the state-of-the-art artificial intelligence and machine learning techniques. These images contain a rich multi-scale information able to represent the amount of topology and its persistence, expected describing the fibers trajectories depending on the concentration and shear rate, the former induing the amount of topological events and the last their persistence.

Figure 9 schematizes the persistence image content, where the horizontal axis refers to the value at which the topological event appears, while the vertical one refers to its persistence. Thus, Figures 10–13 represent the persistence images associated respectively to Figures 1–4, that describe the findings just discussed when referring to the associated persistence diagrams (Figures 5–8).

To sum up the effect of the concentration and the applied shear rate on the persistence image, Figure 14 represents the images corresponding to $\dot{q}_y$ in the different concentration/shear rate conditions, where a clear evolution of the topological pattern can be appreciated.

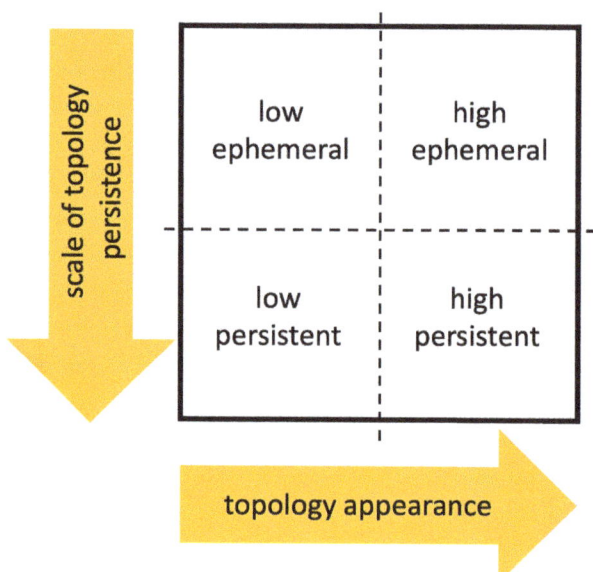

Figure 9. Persistence images reader code.

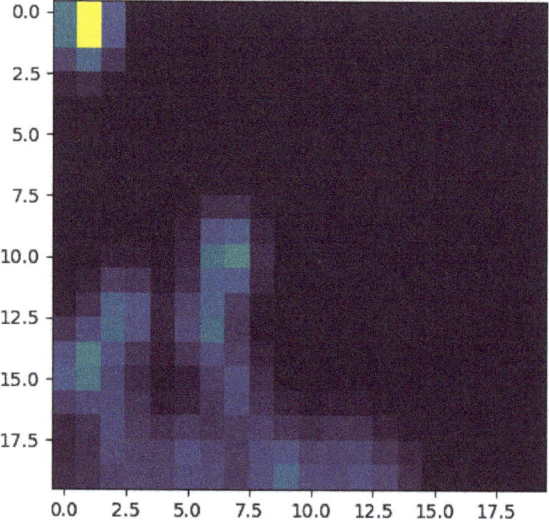

Figure 10. Persistence image related to displacement on the y-direction in Case 4: Minimum shear rate.

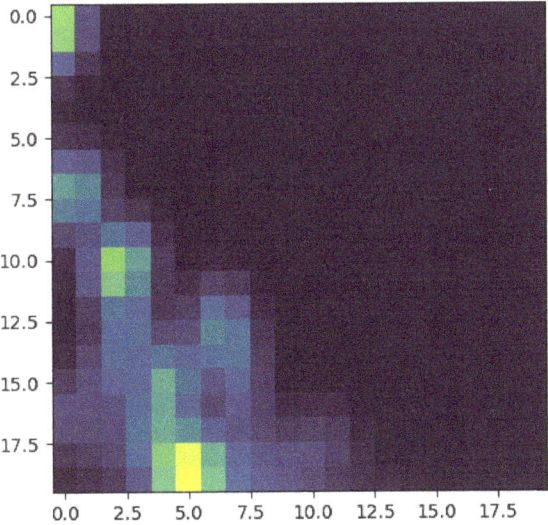

Figure 11. Persistence image related to displacement on the y-direction in Case 6: Maximum shear rate.

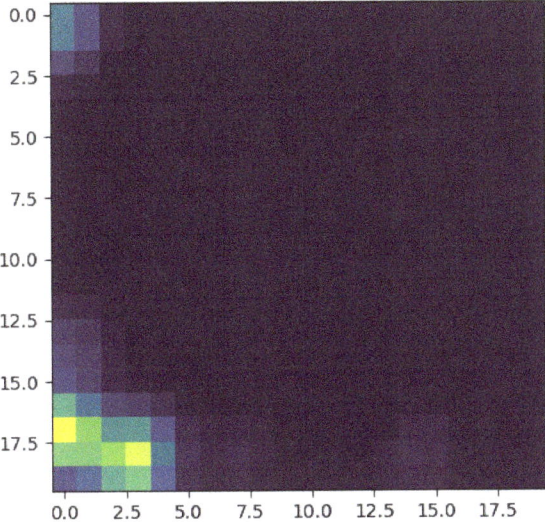

Figure 12. Persistence image related to displacement on the y-direction in Case 1: Minimum concentration.

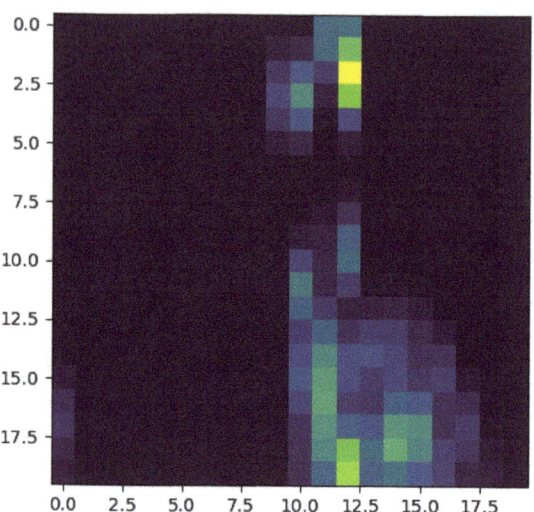

Figure 13. Persistence image related to displacement on the y-direction in Case 3: Maximum concentration.

		Case 1: 5s^{-1}, 14%	
Case 4: 1s^{-1}, 18%	**Case 5:** 3s^{-1}, 18%	**Case 2:** 5s^{-1}, 18%	**Case 6:** 7s^{-1}, 18%
		Case 3: 5s^{-1}, 22%	

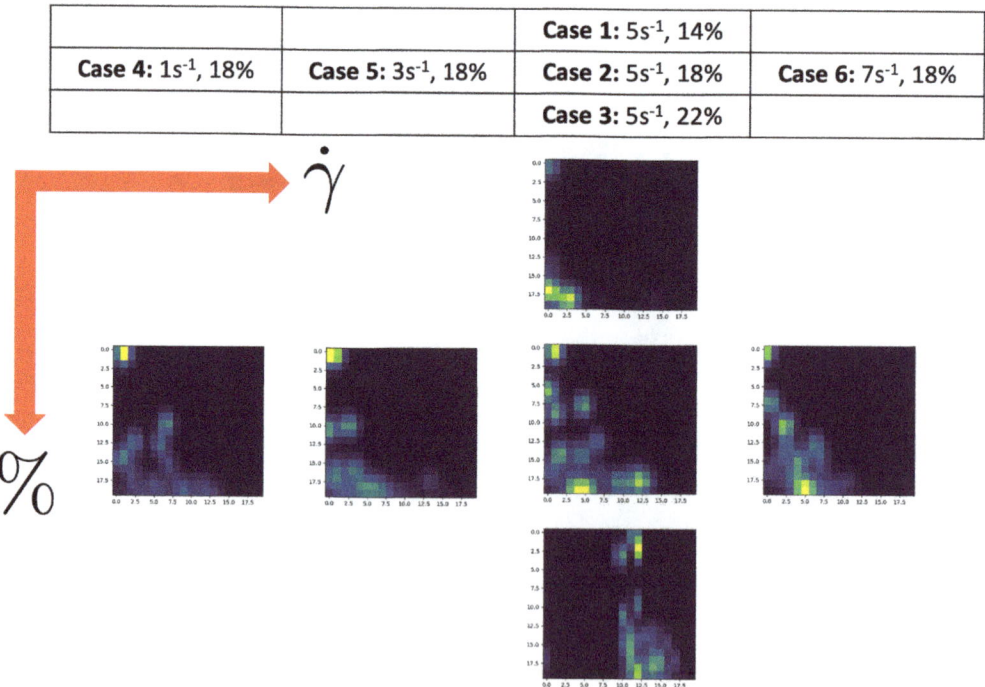

Figure 14. $\dot{q}_y$ persistence images in different fibers volume fraction/shear rate conditions.

5. Conclusions

This paper proved that interactions affect, in a very precise way, the trajectory followed by the geometrical center of the interacting particles. Because of the high variability, a robust metric was chosen for comparison purposes, concretely topological data analysis.

Thus, the time-series related to the erratic perturbation of the nominal trajectories, reflecting the interactions (lubrication and contact) allows to extract a sort of topological pattern, the so-called persistence image, that characterizes in a stable manner (invariant description) all the trajectories related to the same flow conditions, in particular same values of the fiber concentration and flow shear rate, both effecting the number and intensity of the interactions, and then having a noticeable effect on the trajectory topology.

This work opens numerous perspectives, in particular the one related to the flow monitoring, to infer, from the recorded trajectory, local quantities, like the the effective shear rate, concentration and ensemble orientation (moments of the orientation distribution function).

Author Contributions: Conceptualization, F.C.; methodology, R.M., J.A. and N.H.; software, R.M. and N.H. All authors have read and agreed to the published version of the manuscript.

Funding: This research received no external funding.

Institutional Review Board Statement: Not applicable.

Informed Consent Statement: Not applicable.

Data Availability Statement: Data is available under request.

Acknowledgments: Authors acknowledge the CREATE-ID ESI-ENSAM Chair .

Conflicts of Interest: The authors declare no conflict of interest.

References

1. Jeffery, G.B. The motion of ellipsoidal particles immersed in a viscous fluid. *Proc. R. Soc. Lond.* **1922**, *A102*, 161–179.
2. Binetruy, C.; Chinesta, F.; Keunings, R. *Flows in Polymers, Reinforced Polymers and Composites: A Multiscale Approach*; Springerbriefs; Springer: Cham, Switzerland; Heidelberg, Germany; New York, NY, USA; Dordrecht, The Netherlands; London, UK, 2015.
3. Folgar, F.; Tucker, C. Orientation behavior of fibres in concentrated suspensions. *J. Reinf. Plast. Comp.* **1984**, *3*, 98–119. [CrossRef]
4. Advani, S.; Tucker, C. The use of tensors to describe and predict fibre orientation in short fibre composites. *J. Rheol.* **1987**, *31*, 751–784. [CrossRef]
5. Mezher, R.; Abisset-Chavanne, E.; Ferec, J.; Ausias, G.; Chinesta, F. Direct simulation of concentrated fiber suspensions subjected to bending effects. *Model. Simul. Mater. Sci. Eng.* **2015**, *23*, 055007. [CrossRef]
6. Mezher, R.; Perez, M.; Scheuer, A.; Abisset-Chavanne, E.; Chinesta, F.; Keunings, R. Analysis of the Folgar & Tucker model for concentrated fibre suspensions via direct numerical simulation. *Compos. Part A* **2016**, *91*, 388–397.
7. Kim, S.; Karrila, S.J. *Microdynamics: Principles and Selected Applications*; Butterworth-Heinemann: Oxford, UK, 1991.
8. Yamane, Y.; Kaneda, Y.; Doi, M. Numerical-simulation of semidilute suspensions of rodlike particles in shear-flow. *J. Non Newton. Fluid Mech.* **1994**, *54*, 405–421. [CrossRef]
9. Ausias, G.; Fan, X.J.; Tanner, R. Direct simulation for concentrated fibre suspensions in transient and steady state shear flows. *J. Non-Newton. Fluid Mech.* **2006**, *135*, 46–57. [CrossRef]
10. Rabadan, R.; Blumberg, A.J. *Topological Data Analysis For Genomics and Evolution*; Cambridge University Press: Cambridge, UK, 2020.
11. Oudot, S.Y. *Persistence Theory: From Quiver Representation to Data Analysis*; Mathematical Surveys and Monographs; American Mathematical Society: Providence, RI, USA, 2010; Volume 209.
12. Chazal, F.; Michel, B. An introduction to Topological Data Analysis: Fundamental and practical aspects for data scientists. *J. Société Française Stat.* **2017**. Available online: https://arxiv.org/abs/1710.04019 (accessed on 13 September 2021).
13. Peyré, G.; Cuturi, M. Computational Optimal Transport. *Found. Trends Mach. Learn.* **2019**, *11*, 355–607. [CrossRef]
14. Frahi, T.; Chinesta, F.; Falco, A.; Badias, A.; Cueto, E.; Choi, H.Y.; Han, M.; Duval, J.L. Empowering Advanced Driver-Assistance Systems from Topological Data Analysis. *Mathematics* **2021**, *9*, 634. [CrossRef]
15. Frahi, T.; Yun, M.; Argerich, C.; Falco, A.; Chinesta, F. Tape Surfaces Characterization with Persistence Images. *AIMS Mater. Sci.* **2020**, *7*, 364–380. [CrossRef]
16. Frahi, T.; Falco, A.; Vinh Mau, B.; Duval, J.L.; Chinesta, F. Empowering Advanced Parametric Modes Clustering from Topological Data Analysis. *Appl. Sci.* **2021**, *11*, 6554. [CrossRef]
17. Yun, M.; Argerich, C.; Cueto, E.; Duval, J.L.; Chinesta, F. Nonlinear regression operating on microstructures described from Topological Data Analysis for the real-time prediction of effective properties. *Materials* **2020**, *13*, 2335. [CrossRef] [PubMed]

Article

Numerical Prediction of Two-Phase Flow through a Tube Bundle Based on Reduced-Order Model and a Void Fraction Correlation

Claire Dubot [1,2,*], Cyrille Allery [1], Vincent Melot [2], Claudine Béghein [1], Mourad Oulghelou [3] and Clément Bonneau [2]

[1] LaSIE, UMR-7356-CNRS, La Rochelle Université, Avenue Michel Crépeau, 17042 La Rochelle, France; cyrille.allery@univ-lr.fr (C.A.); Claudine.Beghein@univ-lr.fr (C.B.)
[2] Naval Group, Rue du Bac, 44620 La Montagne, France; Vincent.Melot@naval-group.com (V.M.); clement.bonneau@naval-group.com (C.B.)
[3] LAMPA, ENSAM Paris Tech, 2 Boulevard de Ronceray, 49035 Angers, France; mourad.oulghelou@ensam.eu
* Correspondence: Claire.Dubot1@univ-lr.fr

Citation: Dubot, C.; Allery, C.; Melot, V.; Béghein, C.; Oulghelou, M.; Bonneau, C. Numerical Prediction of Two-Phase Flow through a Tube Bundle Based on Reduced-Order Model and a Void Fraction Correlation. *Entropy* **2021**, *23*, 1355. https://doi.org/10.3390/e23101355

Academic Editor: Donald J. Jacobs

Received: 31 August 2021
Accepted: 13 October 2021
Published: 16 October 2021

Publisher's Note: MDPI stays neutral with regard to jurisdictional claims in published maps and institutional affiliations.

Copyright: © 2021 by the authors. Licensee MDPI, Basel, Switzerland. This article is an open access article distributed under the terms and conditions of the Creative Commons Attribution (CC BY) license (https://creativecommons.org/licenses/by/4.0/).

Abstract: Predicting the void fraction of a two-phase flow outside of tubes is essential to evaluate the thermohydraulic behaviour in steam generators. Indeed, it determines two-phase mixture properties and affects two-phase mixture velocity, which enable evaluating the pressure drop of the system. The two-fluid model for the numerical simulation of two-phase flows requires interaction laws between phases which are not known and/or reliable for a flow within a tube bundle. Therefore, the mixture model, for which it is easier to implement suitable correlations for tube bundles, is used. Indeed, by expressing the relative velocity as a function of slip, the void fraction model of Feenstra et al. and Hibiki et al. developed for upward cross-flow through horizontal tube bundles is introduced and compared. With the method suggested in this paper, the physical phenomena that occur in tube bundles are taken into consideration. Moreover, the tube bundle is modelled using a porous media approach where the Darcy–Forchheimer term is usually defined by correlations found in the literature. However, for some tube bundle geometries, these correlations are not available. The second goal of the paper is to quickly compute, in quasi-real-time, this term by a non-intrusive parametric reduced model based on Proper Orthogonal Decomposition. This method, named Bi-CITSGM (Bi-Calibrated Interpolation on the Tangent Subspace of the Grassmann Manifold), consists in interpolating the spatial and temporal bases by ITSGM (Interpolation on the Tangent Subspace of the Grassmann Manifold) in order to define the solution for a new parameter. The two developed methods are validated based on the experimental results obtained by Dowlati et al. for a two-phase cross-flow through a horizontal tube bundle.

Keywords: steam generator; void fraction; mixture model; porous media approach; reduced-order model; Proper Orthogonal Decomposition (POD)

1. Introduction

Steam generators are heat exchangers used especially in nuclear propulsion. Water, heated by the reactor core, flows through a tube bundle, which is a closed circuit called the primary circuit. The heat of the primary fluid is diffused by conduction through metallic tube walls to the water, which flows outside the tubes. Water in the secondary circuit, also called the secondary fluid, enters in a liquid state and becomes a two-phase mixture of steam and water as heat transfer occurs along the heat exchanger. The steam is then used to generate electricity using rotating turbines.

A three-dimensional thermo-hydraulic analysis is essential to predict the performance of heat exchangers and their correct design, especially taking into account that the tube bundle, where there may be thousands of tubes, would require unacceptable computational cost and time. Therefore, the whole code cited in this paper models the tube bundle as a porous medium. In the case of two-phase flows, the void fraction is the key parameter to characterize the flow. Indeed, it enables calculating the mixture density, the mixture viscosity, and the mixture velocity. Consequently, it plays an important role for computing pressure drops and heat and mass transfers.

Since the 1980s, thermal-hydraulic codes have been developed to understand the physical phenomena involved. One of the first codes, THIRST [1], was developed to compute three-dimensional, two-phase, and steady flow in steam generators. The two-phase flow was solved by the homogeneous two-phase model, and the phase velocities were assumed to be equal, but this does not reflect what is really going on. To take into account the slip between phases, Navier–Stokes equations were solved for the mixture of the secondary fluid. The THYC code [2] gives the relative velocity thanks to a correlation using the drift flux model of Zuber and Findlay [3]. This model is based on the determination of drift-flux parameters for which there are many different empirical correlations [4–8]. Stevanovic et al. [9] used the two-fluid model to predict the thermal-hydraulic behavior in horizontal tube bundles. Navier–Stokes equations were solved for each phase. The accurate definition of the interfacial drag force, comprising a drag coefficient correlation, is important in order to predict the void fraction distribution. In their paper, the original drag correlation of Ishii and Zuber [10] was multiplied by 0.4. They validated this modification of the correlation with their experimental data. Nevertheless, most drag coefficient laws in the literature [11–15], including Ishii and Zuber's correlation, are made for different two-phase flow regimes (bubbly, slug, stratified, annular, or spray flow) inside a tube and not in tube bundles. In this study, this problem is tackled by using the mixture model, which is a simplified two-phase model where Navier–Stokes equations are solved for the mixture. The developed method involved formulating the relative velocity as a function of slip and then implementing a specific void fraction model for tube bundles derived from the literature.

The slip ratio is defined as the ratio between gas phase velocity and liquid phase velocity. Feenstra et al. [16] developed a slip ratio model based on their R-11 data for upward two-phase cross-flow through horizontal tube bundles. They identified the important variables that affected slip, and the application of the Buckingham pi theorem enabled them to reduce the slip ratio as a function of two dimensionless numbers, namely the Richardson number and the Capillary number. They demonstrated that it fitted well with experimental void fraction data in R-11 and air–water mixtures for a wide range of mass fluxes, qualities and pitch-diameter ratios. This model is not explicit for void fraction; indeed an iterative process is necessary to compute the void fraction. Likewise, Hibiki et al. [17] used the slip ratio correlation of Smith [18] to develop a correlation for the entrainment factor dependent on the tube's bundle arrangement. The entrainment factor is correlated with a dominant parameter such as nondimensional mass flux based on experimental data coming from various flow configurations and tube bundle arrangements. The developed correlation agrees both with parallel and crossflow in the tube bundle.

Moreover, the porous media approach implies adding a momentum sink to the governing momentum equation. This source term is defined by the Darcy–Forchheimer law [19,20]. A widely used and easy method is the use of correlations coming from the literature, such as Zukauskas et al.'s correlation [21] for a transverse flow to a tube bundle or Rhodes and Carlucci's correlation [22] for a parallel flow. These correlations result from experiments, it seems to be the most accurate method to determine the law but it is valid for a given design and it is expensive and time-consuming. However, for new, i.e., non-standard, tube geometry, these correlations are not available, and solving a reduced-order model to define them is suggested. From some CFD calculations of flow through a Representative Elementary Volume (REV) of the tube bundle, pressure and velocity fields were decomposed by POD (Proper Orthogonal Decomposition). Then, the non-intrusive reduced model method, Bi-CITSGM (Bi-Calibrated Interpolation on the Tangent Subspace of the Grassmann Manifold) [23,24], was applied in order to determine the solution for a new parameter. In this method, spatial and temporal POD sampling bases are interpolated by ITSGM (Interpolation on the Tangent Subspace of the Grassmann Manifold) [25] and the temporal eigenvalues by usual methods such as Lagrange, IDW (Inverse Distance Weighting), or RBF (Radial Basis Function). Then, the reduced-order model was used to compute the source term of the porous media approach applied to the flow through a tube bundle.

This work was validated with the experimental results of Dowlati et al. [26] which is a two-phase cross-flow through a horizontal tube bundle. The summary of the developed methodology is presented in Figure 1. First, the mixture model was used, and we rewrote the slip velocity, $\vec{u}_{gl}$, as a function of the slip in order to implement a specific void fraction model. The implementation of Feenstra's correlation and Hibiki's correlation were compared to the usual formulation of Manninen et al. [27]. Moreover, the tube bundle was modelled by a porous medium. The ability to use a reduced-order model to determine the momentum sink term $\vec{F}_t$ was studied and compared to the usual correlation of Zukauskas et al. [21].

The remainder of the paper is organized as follows. In Section 2, the governing equations used to model the two-phase flow through a tube bundle by a porous media approach are presented. The rewriting of the relative velocity and the Darcy–Forchheimer term are also detailed. Section 3 deals with the methodology of the reduced-order model on the REV to compute the Forchheimer term. The ITSGM method and the non-intrusive approach Bi-CITSGM are reviewed. Section 4 validates the application of the proposed reformulation of the relative velocity, and Section 5 confirms the use of a ROM to determine the momentum sink term. Finally, conclusions are drawn.

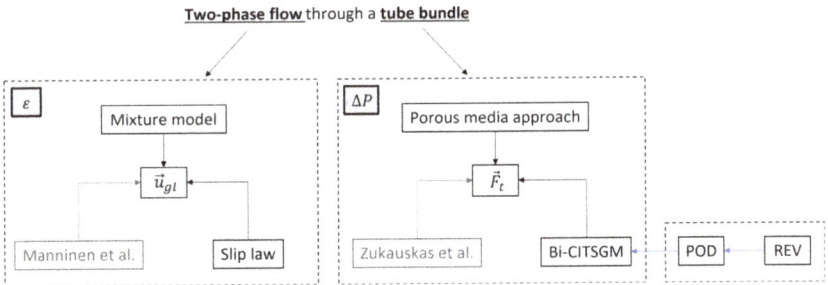

Figure 1. Summary of the suggested approach. On the left side, the mixture model is modified by Feenstra's correlation to compute the void fraction. On the right side, the Bi-CITSGM method applied to a REV is introduced in the porous media approach to define the pressure drop of the tube bundle.

2. Cross-Flow through a Horizontal Tube Bundle

This study focuses on the simulation of the thermal-hydraulic behaviour of a two-phase adiabatic crossflow through a horizontal tube bundle. The tube bundle is modelled by a porous medium that involves solving the conservation equations of mass and momentum and the turbulence model using the superficial velocity porous formulation. The flow is considered two-dimensional and only transverse to the tube bundle. All geometric notations pertaining to the tube bundle are illustrated in Figure 2.

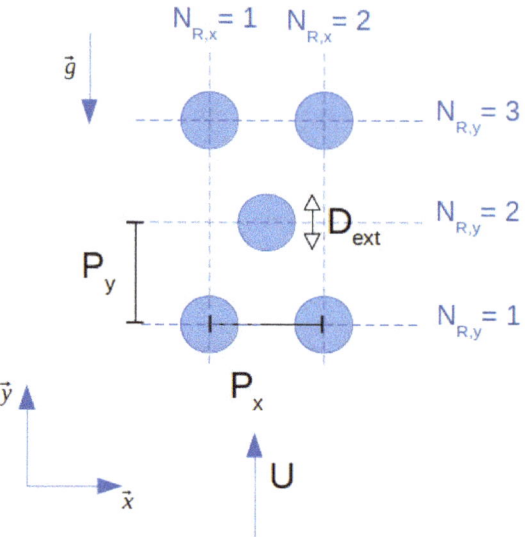

Figure 2. Geometric definitions of the tube bundle.

2.1. Governing Equations of the Two-Phase Flow

In order to simulate the adiabatic two-phase flow, the mixture model is used. Only two phases are considered: the liquid phase l and the gas phase g. The mixture model allows the phases to be interpenetrating and to move at different velocities using the concept of slip velocities. The continuity equation for the mixture is

$$\frac{\partial}{\partial t}(\rho_b) + \nabla \cdot (\rho_b \vec{u}_b) = 0 \qquad (1)$$

where $\vec{u}_b$ is the mass-averaged velocity defined by

$$\vec{u}_b = \frac{\alpha_g \rho_g \vec{u}_g + \alpha_l \rho_l \vec{u}_l}{\rho_b}. \qquad (2)$$

α_k is the volume fraction of phase k, and ρ_b is the mixture density, defined by

$$\rho_b = \alpha_g \rho_g + \alpha_l \rho_l. \qquad (3)$$

The momentum equation for the mixture is obtained by summing the individual momentum equations of all phases. It takes the following form

$$\frac{\partial}{\partial t}(\rho_b \vec{u}_b) + \nabla \cdot (\rho_b \vec{u}_b \otimes \vec{u}_b) = -\nabla p + \nabla \cdot \left[\mu_b \left(\nabla \vec{u}_b + \nabla \vec{u}_b^T\right)\right] + \rho_b \vec{g} \\ - \nabla \cdot \left(\alpha_g \rho_g \vec{u}_{dr,g} \vec{u}_{dr,g} + \alpha_l \rho_l \vec{u}_{dr,l} \vec{u}_{dr,l}\right) + \vec{F}_t. \qquad (4)$$

μ_b is the mixture viscosity such as

$$\mu_b = \alpha_g \mu_g + \alpha_l \mu_l \qquad (5)$$

and $\vec{g} = -g\vec{y}$ with $g = 9.81$ m/s^2. The hydraulic resistance of tubes on the fluid is taken into account by the Darcy–Forchheimer term $\vec{F}_t$. This is a source term due to the use of the porous media approach and it is defined in the following Section 2.3. $\vec{u}_{dr,k}$ is the drift velocity of the phase k defined by

$$\vec{u}_{dr,k} = \vec{u}_k - \vec{u}_b. \tag{6}$$

Moreover, the volume fraction equation for phase g is built from the continuity equation for the gas phase, using the definition of the drift velocity (Equation (6)) to eliminate the phase velocity as:

$$\frac{\partial}{\partial t}\left(\alpha_g \rho_g\right) + \nabla \cdot \left(\alpha_g \rho_g \vec{u}_b\right) = -\nabla \cdot \left(\alpha_g \rho_g \vec{u}_{dr,g}\right). \tag{7}$$

The drift velocity is related to the relative (or slip) velocity according to:

$$\vec{u}_{dr,g} = \vec{u}_{gl}\left(1 - \frac{\alpha_g \rho_g}{\rho_b}\right) \quad \text{and} \quad \vec{u}_{dr,l} = \vec{u}_{gl}\left(\frac{\alpha_l \rho_l}{\rho_b} - 1\right). \tag{8}$$

The most used algebraic slip formulation is the Manninen model [27]. With this formulation, the slip velocity is given by

$$\vec{u}_{gl} = \frac{\tau_g}{f_{drag}} \frac{(\rho_g - \rho_b)}{\rho_g} \vec{a} \tag{9}$$

where τ_g is the particle relaxation time,

$$\tau_g = \frac{\rho_g d_g^2}{18 \mu_g} \tag{10}$$

d_g is the gas particle diameter and $\vec{a}$ is the acceleration.

$$\vec{a} = \vec{g} - (\vec{u}_b \cdot \nabla)\vec{u}_b - \frac{\partial \vec{u}_b}{\partial t} \tag{11}$$

The drag force f_{drag} is obtained with the model of Schiller and Naumann [11]. Commonly used, it is expressed as a function of the drag coefficient C_D and the relative Reynolds number:

$$f_{drag} = \frac{C_D Re}{24} \quad \text{and} \quad Re = \frac{\rho_l \|\vec{u}_g - \vec{u}_l\| d_g}{\mu_l} \tag{12}$$

This slip velocity formulation is not suitable for a two-phase flow in tube bundles because it is not designed for such a configuration and does not take into account the associated physical phenomena. In the next subsection, a formulation more adequate for the tube bundle configuration is suggested.

2.2. Rewriting of the Slip Velocity

The void fraction depends on the slip ratio S and quality x according to

$$\varepsilon = \frac{1}{1 + S \frac{\rho_g}{\rho_l} \frac{1-x}{x}} \tag{13}$$

Here, the slip velocity is reformulated as a function of slip in order to introduce a slip ratio model adapted to a two-phase through a tube bundle. The slip velocity is the velocity difference between the gas phase and the liquid phase.

$$\vec{u}_{gl} = \vec{u}_g - \vec{u}_l \tag{14}$$

The velocity components in Equation (14) can be written as

$$\begin{cases} u_{gl,x} = u_{l,x}\left(\frac{u_{g,x}}{u_{l,x}} - 1\right) = u_{l,x}(S_x - 1) \\ u_{gl,y} = u_{l,y}\left(\frac{u_{g,y}}{u_{l,y}} - 1\right) = u_{l,y}(S_y - 1) \end{cases} \quad (15)$$

$u_{l,x}$ (resp. $u_{l,y}$) is the liquid velocity in the $\vec{x}$ (resp. $\vec{y}$) direction. The same notation is used for the gas velocity and the slip velocity.

For an upward cross-flow to the tube bundle in the y direction, it can be assumed that the velocity ratio S_x is equal to 1 and that S_y is defined by a correlation coming from the literature. Finally, the drift velocity is written as

$$\begin{cases} u_{dr,g,x} = u_{l,x}(S_x - 1)\left(1 - \frac{\alpha_g \rho_g}{\rho_b}\right) \\ u_{dr,g,y} = u_{l,y}(S_y - 1)\left(1 - \frac{\alpha_g \rho_g}{\rho_b}\right) \end{cases} \quad (16)$$

The proposed approach needs to solve Equations (1), (4) and (7) with the modified definition of the drift velocity defined by Equation (16).

2.2.1. Hibiki's Correlation (2017)

Smith [18] gives the slip ratio valid for all void fraction ranges and for vertical, inclined, and horizontal flows in a channel [28,29] as

$$S = e + (1 - e)\left(\frac{\frac{\rho_l}{\rho_g} + \left(\frac{1-x}{x}\right)e}{1 + \left(\frac{1-x}{x}\right)e}\right)^{0.5} \quad (17)$$

where e is the ratio of the mass of liquid droplets entrained in the gas core to the total mass of liquid. In order to implement a void fraction correlation in a steam generator thermal-hydraulic code, Hibiki et al. defined this parameter for staggered tube bundles like Dowlati et al. as

$$e = min(0.0637 N_{\dot{m},p}^{0.571}, 1). \quad (18)$$

$N_{\dot{m},p} = \dot{m}_p / \rho_g j_{g,crit}$ where the critical superficial gas velocity is

$$j_{g,crit} = \left(\frac{\Delta \rho g \sigma}{\rho_g^2}\right)^{0.25} \left(\frac{\mu_l}{\left(\rho_l \sigma \sqrt{\frac{\sigma}{g\Delta\rho}}\right)^{0.5}}\right)^{-0.2}. \quad (19)$$

where σ is the surface tension. $\dot{m}_p$ is the pitch mass flux, which represents the mixture velocity between two tubes $\|\vec{u}_{b,p}\|$ multiplied by the mixture density.

$$\dot{m}_p = \alpha_g \rho_g \|\vec{u}_{g,p}\| + \alpha_l \rho_l \|\vec{u}_{l,p}\| \quad \text{where} \quad \|\vec{u}_{k,p}\| = \|\vec{u}_k\| \frac{P}{P - D_{ext}} \quad \text{for } k = \{l, g\} \quad (20)$$

2.2.2. Feenstra's Correlation (2000)

Feenstra et al. [16] defined the slip ratio as

$$S = 1 + 25.7\sqrt{Ri * Cap}\left(\frac{P}{D_{ext}}\right)^{-1} \quad (21)$$

D_{ext} is the outer diameter of the tubes, and P is the pitch, illustrated in Figure 2. The Richardson number is the ratio between the buoyancy force and the inertia force:

$$Ri = \frac{(\rho_g - \rho_l)^2 g(P - D_{ext})}{\dot{m}_p^2} \quad (22)$$

The Capillary number is the ratio between the viscous force and the surface tension force:

$$Cap = \frac{\mu_l \|\vec{u}_{g,p}\|}{\sigma} \tag{23}$$

The gas phase velocity is based on the resulting void fraction:

$$\|\vec{u}_{g,p}\| = \frac{x \dot{m}_p}{\varepsilon \rho_g} \tag{24}$$

2.3. Definition of the Darcy–Forchheimer Term

The source term $\vec{F}_t = F_{t,x}\vec{x} + F_{t,y}\vec{y}$ is added to the momentum equation because tube bundles are represented by a porous medium and is usually written as:

$$\begin{cases} F_{t,x} = -\left(D_{xx} \mu_b u_{b,x} + K_{xx} \frac{1}{2} \rho_b \|\vec{u}_b\| u_{b,x} \right) \\ F_{t,y} = -\left(D_{yy} \mu_b u_{b,y} + K_{yy} \frac{1}{2} \rho_b \|\vec{u}_b\| u_{b,y} \right) \end{cases} \tag{25}$$

This term is composed of a viscous loss term and an inertial loss term resulting from Darcy–Forchheimer's law [19]. $\|\vec{u}_b\|$ is the mixture velocity magnitude, D_{xx} and D_{yy} are the inverse of the permeability, and K_{xx} and K_{yy} are the correction terms of Forchheimer. In this study, the first term of Equation (25) is neglected because the flow is turbulent. As the term has the same dimensions as a pressure gradient, Equation (25) is rewritten by:

$$F_{t,x} = -\frac{\Delta P_{f,x}^{2\Phi}}{N_{R,x} P_x} \quad \text{and} \quad F_{t,y} = -\frac{\Delta P_{f,y}^{2\Phi}}{N_{R,y} P_y} \tag{26}$$

$\Delta P_{f,i}^{2\Phi}$ is the two-phase frictional pressure drop, $N_{R,i}$ is the number of tube rows, and P_i is the pitch in direction i ($\vec{x}$ or $\vec{y}$) shown in Figure 2. From Equations (25) and 26, the unknown coefficients K_{xx} and K_{yy} are defined by:

$$\begin{cases} K_{xx} = -\frac{\Delta P_{f,x}^{2\Phi}}{\frac{1}{2}\rho_b \|\vec{u}_b\| u_{b,x} N_{R,x}} \frac{1}{P_x} \\ K_{yy} = -\frac{\Delta P_{f,y}^{2\Phi}}{\frac{1}{2}\rho_b \|\vec{u}_b\| u_{b,y} N_{R,y}} \frac{1}{P_y} \end{cases} \tag{27}$$

From the method developed by Consolini et al. [30] to define the two-phase frictional pressure drop over horizontal tube bundles, Equation (27) is reduced to:

$$K_{xx} = -\frac{\lambda Eu}{P_x} \left(\frac{P_y}{P_y - D_{ext}} \right)^2 \quad \text{and} \quad K_{yy} = -\frac{\lambda Eu}{P_y} \left(\frac{P_x}{P_x - D_{ext}} \right)^2 \tag{28}$$

where the Euler number, Eu, can be given by Zukauskas et al. [21]. Zukauskas et al. defined a correlation for the Euler number resulting from their experiments and experimental results from the literature. This law enables to determine frictional pressure drops for in-line and staggered tube bundles with $1.25 \leq P/D_{ext} \leq 2.5$ and $10 \leq Re \leq 10^6$. The Euler number is calculated as:

$$Eu = k_1 \sum_{i=0}^{4} \frac{c_i}{Re^i} \tag{29}$$

where c_i and k_1 are coefficients given in reference [21]. The two-phase multiplier coefficient is written by Consolini et al. [30] as:

$$\lambda = \Lambda + (1 - \Lambda)(1 - 2x)^2 \quad \text{with} \quad \Lambda = \left(\frac{\dot{m}_p}{400} \right)^{-1.5} \tag{30}$$

It is important to note that the two-phase multiplier factor is equal to 1 when the quality tends towards 0 (only liquid phase) and 1 (only gas phase). This key argument is not available with the correlation of Ishihara et al. [31] based on the Lockhart–Martinelli approach [32], which was the method used by Dowlati et al. in [26].

For complex or non-standard geometries, such as helicoidal tube bundles or corrugated tubes [33–35], there are no suitable or reliable correlations in the literature to compute the Forchheimer force $\vec{F}_t$. They can be obtained by experiments or numerical simulations by LES or RANS approaches but these methods imply a significant cost. In order to reduce computational time and cost, we suggest using a non-intrusive reduced model model (ROM). This parametric ROM simulates the flow in an REV (Representative Elementary Volume) of the tube bundle. The knowledge of this flow enables the Forchheimer term to be quickly obtained. In addition, the spatial distributions of the pressure and the velocity around each tube are precisely given by this approach. The ROM methodology is detailed in the next section.

3. Reduced Ordel Model on the REV to Compute the Forchheimer Term

3.1. Representative Elementary Volume

A Representative Elementary Volume (REV) of the tube bundle is considered in Figure 3, where the lower and upper boundary conditions are periodical. Governing equations for periodically fully developed flow are derived from the incompressible Navier–Stokes equations defined by:

$$\begin{cases} \nabla \cdot \vec{u} = 0 \\ \frac{\partial \vec{u}}{\partial t} + \nabla \cdot (\vec{u} \otimes \vec{u}) = -\frac{\nabla P}{\rho} + \nu \nabla^2 \vec{u} + \frac{\vec{\beta}}{\rho} \end{cases} \quad \text{where} \quad \vec{\beta} = \begin{pmatrix} 0 \\ \beta \end{pmatrix} \quad (31)$$

$\vec{u}$ is the periodic flow velocity:

$$\vec{u}(x, y, t) = \vec{u}(x, y + 2P_y, t) \quad (32)$$

P represents the reduced pressure, which satisfies periodic boundary conditions,

$$P(x, y, t) = P(x, y + 2P_y, t) \quad (33)$$

and the actual pressure is given by

$$p(x, y, t) = -\beta y + P(x, y, t) \quad (34)$$

according to Patankar [36]. β is the linear component of the pressure, which is to be calculated iteratively for a fixed mass flow rate [37,38].

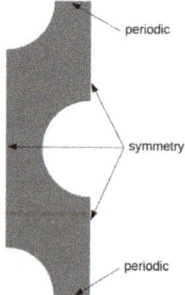

Figure 3. REV of the tube bundle of Dowlati et al. [26].

3.2. Definition of Reduced Bases

Model reduction techniques make it possible to quickly and inexpensively obtain the temporal dynamics of a complex flow. The principle of the reduced-order model (ROM) is to approximate the solution in a small dimension sub-vector space, which enables capturing the dominant characteristics of the physical phenomenon studied. The solution $f(t,x)$ is written as a linear combination of a finite number of spatial basis functions $\Phi^k(x)$ as:

$$f(t,x) \approx \sum_{k=1}^{q} a^k(t) \Phi^k(x) \qquad (35)$$

q is relatively small compared to the problem size, and a^k is the temporal coefficient. Here, f corresponds to the velocity $\vec{u}$ or the pressure p.

The most common method to compute the spatial basis Φ is the POD (Proper Orthogonal Decomposition) method [39,40]. However, the ability of the POD basis function to give the dynamic of the phenomenon studied is dependent of the information contained in the snapshots that form the basis. For instance, a POD basis built with snapshots for a Reynolds number Re_1 will not be able to predict the dynamics of the physical phenomenon for another Reynolds number Re_2. To increase the validity domain of the POD basis, it is possible to interpolate a set of POD bases $\Phi_1, \cdots, \Phi_N$ built for different values of Reynolds numbers $Re_1, \cdots, Re_N$ in order to obtain the basis associated to the desired Reynolds number. Standard interpolation techniques (RBF, Lagrange, Spline, etc.) are not very efficient and are not generally representative of the phenomenon studied. To get around this difficulty, a basis interpolation approach based on the results of differential geometry and, more particularly, on the properties of the Grassmann manifold can be used [25,41–43]. In this work, we consider the approach offered by Amsallem et al. [25] that is subsequently noted as ITSGM (Interpolation on the Tangent Subspace of the Grassmann Manifold). The algorithm is given in Algorithm A1.

Once the spatial basis for the desired parameter is defined, the temporal coefficients are usually computed by solving a system of differential equations (ROM) resulting from the Galerkin projection of the full model on the basis functions $\Phi^k(x)$ [44]. This method has the disadvantage of being costly and intrusive. Indeed, for each new value of Reynolds number, it is necessary to compute the coefficients of the ROM, which is costly. Moreover, derivative operators are difficult to assess using a commercial CFD code. Consequently, in this paper, we use the non-intrusive method Bi-CITSGM [23,45].

3.3. Description of the Bi-CITSGM Method

The methodology of the Bi-CITSGM is given in Algorithm A2 in the Appendix B. The first step of this method is to interpolate the spatial basis and the temporal basis, built by POD, using the ITSGM method. The singular value matrix is acquired by classical interpolation methods such as Lagrange, RBF, and Spline. Then, the ranking step aims to sort the interpolated spatial and time eigen modes with respect to the interpolated singular values. Indeed, as the spatial and temporal bases may not be in the same order as the singular values, orthogonal matrices need to be introduced. These calibration matrices are a solution to an optimization problem under constraints whose solution is analytically determined. For more details, see [23,24].

4. Validation of the Using of the Modified Mixture Model on the Dowlati's Experiment

4.1. Study Configuration

The methodology, presented in Section 2, was validated with the experiment of Dowlati et al. They made void fraction and friction pressure drop measurements for vertical two-phase flow of air–water across staggered in-line tube bundles with different pitch-to-diameter ratios. Here, the tube bundle, illustrated in Figure 4, is made up of 20 tube rows in a staggered arrangement with five tubes in each row and the ratio P/D_{ext} is 1.75. Geometry dimensions are given in Table 1. The estimated uncertainties in the data done by Dowlati et al. are detailed in Table 2. The present study is done with quality range between 1.3×10^{-4} and 3×10^{-2} and mass flux range between 164 kg/(m²· s) and 538 kg/(m²· s).

Table 1. Dimensions of the tube bundle.

	Value
Outer diameter of a tube D_{ext} [mm]	12.7
Ratio P/D_{ext} [-]	1.75
Pitch P_x [mm]	22.225
Pitch P_y [mm]	19.25
Width $L = 5P_x$ [mm]	111.125
Height $H = 20P_y$ [mm]	385

Table 2. Estimated uncertainties in the data given by Dowlati et al. [26].

Parameter	Uncertainty
Quality	±2%
Void fraction	±0.05
Mass flux	±2%
Frictional two-phase pressure drop	±15%

CFD calculations were performed with Ansys Fluent v2020R2. The tubes in the tube bundle were not represented. The tube bundle was modelled as a porous medium by adding a source term in the momentum equation. At the inlet, the homogeneous void fraction model is assumed. Thus, the homogeneous void fraction ε_H and inlet phase velocities are written as follows,

$$\vec{u}_{g,in} = \begin{pmatrix} 0 \\ \frac{x \dot{m}_{in}}{\rho_g \varepsilon_H} \end{pmatrix}_{(\vec{x},\vec{y})} \quad ; \quad \vec{u}_{l,in} = \begin{pmatrix} 0 \\ \frac{(1-x)\dot{m}_{in}}{\rho_l (1-\varepsilon_H)} \end{pmatrix}_{(\vec{x},\vec{y})} \qquad (36)$$

$$\varepsilon_H = \frac{1}{1 + \frac{\rho_g}{\rho_l} \frac{1-x}{x}} \qquad (37)$$

where x is the quality. Calculations are initialized from the input boundary conditions. The mesh, illustrated in Figure 5, is defined in such a way as to ensure that the dimensionless wall distance y^+ is close to 1. The turbulent model k-ω SST [46] is used.

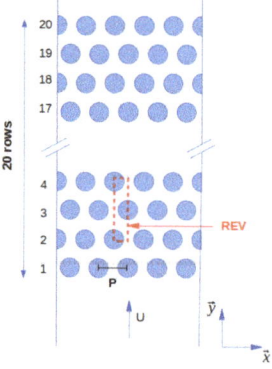

Figure 4. Staggered tube bundle from Dowlati's experiment ($P/D_{ext} = 1.75$).

Figure 5. Mesh used for CFD calculations (size for the central cells = 3 mm/cell size along the border = 0.75 mm).

4.2. Influence of the Slip Model on the Void Fraction Prediction

In order to evaluate the influence of the slip model on the void fraction prediction, mean relative and absolute errors are introduced.

$$E_{rel} = \frac{1}{N_{pt}} \sum_{i=1}^{N_{pt}} \frac{\bar{\varepsilon}_i^{calc} - \bar{\varepsilon}_i^{exp}}{\bar{\varepsilon}_i^{exp}} \tag{38}$$

$$E_{abs} = \frac{1}{N_{pt}} \sum_{i=1}^{N_{pt}} \left(\bar{\varepsilon}_i^{calc} - \bar{\varepsilon}_i^{exp} \right) \tag{39}$$

N_{pt}, $\bar{\varepsilon}^{calc}$, and $\bar{\varepsilon}^{exp}$ are, respectively, the number of data and the computed and experimental void fractions averaged over the tube bundle. To prove that we need to apply a void fraction model to the mixture model, calculations were performed with the slip velocity formulation of Manninen. Here, the value of the void fraction postprocessed with this formulation always matches with the homogeneous void fraction. Figures 6 and 7, for, respectively, $\dot{m}_p = 164$ kg/(m².s) and $\dot{m}_p = 401$ kg/(m².s), showed that the homogeneous void fraction, and thus the Manninen's formulation, do not take into account the slip between phases for flows in tube bundle. Indeed, the relative error is about $E_{rel} = 51.05\%$ and the absolute error $E_{abs} = 0.15$. The results stress that it is important to implement a void fraction model appropriate for a two-phase cross-flow through a horizontal tube bundle. Moreover, for each mass flux, Figures 6 and 7 demonstrate the correct implementation of Feenstra or Hibiki's correlation in the CFD code. That is to say, the errors calculated subsequently in this paper derive only from the slip ratio model implemented and not from the numerical model. It can be seen in Figure 8, which depicts void fraction results as functions of quality and mass flux, that the void fraction increases with the quality and with the mass flux. Roser [47] justified this phenomenon by the upward movement of the gas phase against the liquid phase due to the buoyancy force making it all the more important that the mass flux is low. For higher mass flux, the gap with the homogeneous void fraction model is less significant. Indeed, the two phases are "well mixed" due to the increase in turbulence.

Figure 9 depicts CFD-computed void fractions versus experimental void fractions for all mass fluxes. Feenstra's correlation always underpredicts the void fraction; however, errors are acceptable with $E_{rel} = 17.49\%$ and $E_{abs} = 0.07$. Errors seem to be more important for high mass fluxes and low qualities with some errors outside of range ±20%. Hibiki's correlation overpredicts the results for $\varepsilon \leq 0.5$ and underpredicts it for $\varepsilon > 0.5$. The relative error is a little higher than Feenstra's correlation with $E_{rel} = 21.81\%$, but the absolute error is better with $E_{abs} = 0.05$. Some void fraction points are outside the range ±20% for low void fractions; however, this correlation is better for higher void fractions.

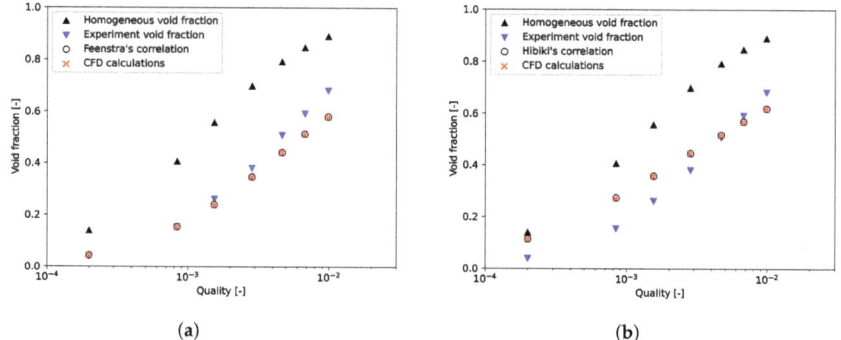

Figure 6. Void fraction results for air-water flow through Dowlati's staggered tube bundle with $\dot{m}_p = 164$ kg/(m².s). (**a**) Feenstra's correlation; (**b**) Hibiki's correlation.

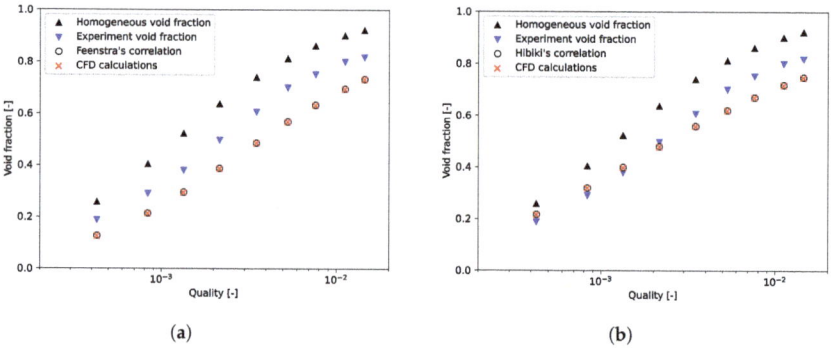

Figure 7. Void fraction results for air-water flow through Dowlati's tube bundle with $\dot{m}_p = 401$ kg/(m².s). (**a**) Feenstra's correlation; (**b**) Hibiki's correlation.

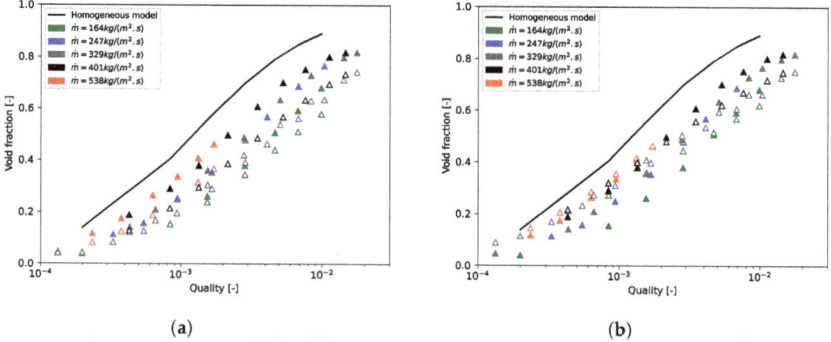

Figure 8. Void fraction as a function of quality and mass flux (△ = results from CFD simulations and ▲ = experimental measurements). (**a**) Feenstra's correlation; (**b**) Hibiki's correlation.

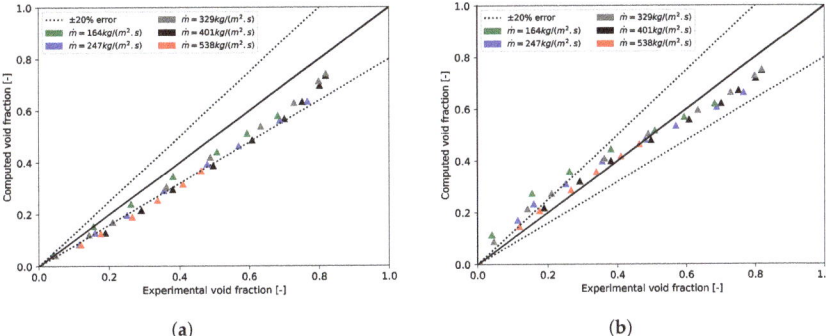

Figure 9. CFD computed void fraction versus experimental void fraction. (**a**) Feenstra's correlation; (**b**) Hibiki's correlation.

In order to improve the void fraction prediction, the Capillary number used in Feenstra's correlation is now expressed as a function of upstream mass flux. The upstream mass flux and the mass flux between two tubes are linked by:

$$\dot{m}_p = \dot{m}_{upstream} \frac{P}{P - D_{ext}} \tag{40}$$

With this modified Feenstra's correlation, named "Upstream Feenstra's correlation", the results were always improved compared to the experimental results, the original Feenstra's correlation, and Hibiki's correlation. Indeed, Figure 10 shows that the results with the modified Feenstra's correlation were closer to experimental results than the original Feenstra's correlation. It is important to note that only 10% of the simulations had a relative error for the void fraction higher than 20% for the modified Feenstra's correlation. As can be seen in Figure 11, these points were located at low void fractions. For the higher void fraction, results were very close to the experiment. Compared to other correlations, this one has an absolute error of 0.03 and a relative error of 9.10%, which confirms the accuracy of the void fraction prediction.

Figure 12 plots the slip ratio obtained by the experiment and Feenstra, Hibiki and upstream Feenstra's correlations as a function of quality for each mass flux. As the void fraction increases with mass flux, it can be noticed that slip ratio decreases with the quality until reaching 1. For $\dot{m}_p \leq 247$ kg/(m².s), the slip ratio turns out to be relatively constant regardless of the quality. However, no correlation studied here captures this phenomenon. For $\dot{m}_p \geq 329$ kg/(m².s), the slip ratio increases with the quality, and correlations follow the same trend. Overall, the slip ratio obtained by the upstream Feenstra correlation is the closest to experimental results.

Now that we have verified the accurate prediction of the void fraction by rewriting the relative velocity and implementing a suitable void fraction model, we can focus on the pressure drop of the system. The porous media approach implies adding a source term to the momentum equation. We prove, in the next section, the possibility to use a non-intrusive parametric reduced model on an REV in order to compute this term.

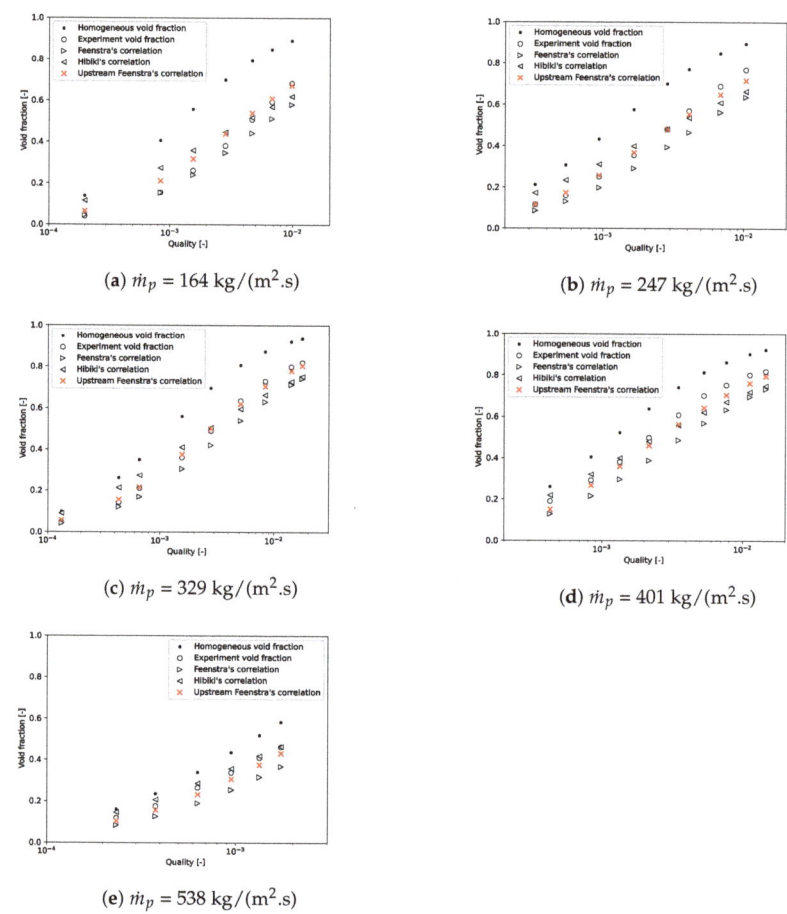

(a) $\dot{m}_p = 164$ kg/(m^2.s)

(b) $\dot{m}_p = 247$ kg/(m^2.s)

(c) $\dot{m}_p = 329$ kg/(m^2.s)

(d) $\dot{m}_p = 401$ kg/(m^2.s)

(e) $\dot{m}_p = 538$ kg/(m^2.s)

Figure 10. Comparison of void fraction results obtained by the different methods.

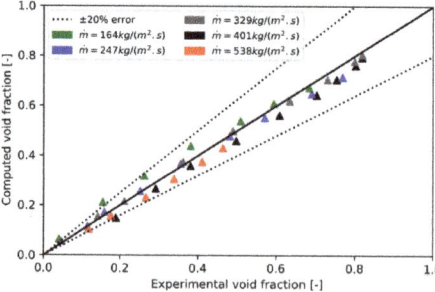

Figure 11. CFD computed void fraction with the modified Feenstra's correlation versus experimental void fraction.

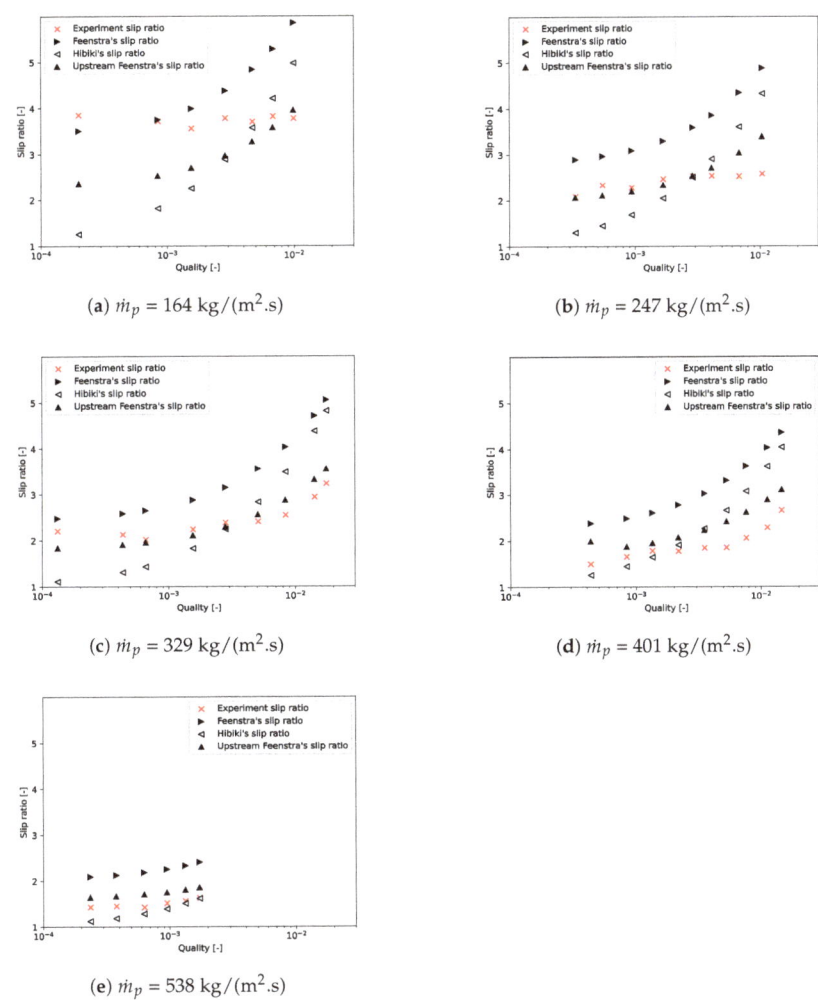

Figure 12. Comparison of slip ratio obtained by the different methods.

5. Computation of the Pressure Drop in Tube Bundles by Using POD-ROM on an REV

The porous media approach involves the implementation of the Euler number (Equation (28)) in the Forchheimer's correction term. Usually, this dimensionless number is computed with correlations coming from the literature. Here, we want to show that it is possible to compute this variable by a non-intrusive reduced-order model, Bi-CITSGM, applied to a REV of the tube bundle. The case of the tube bundle of Dowlati et al. is an case of application for which there is already a correlation resulting from the literature and for which the resolution of a reduced model is of little interest. However, we would like to extend this developed methodology to a case of a complex tube bundle for which there is no correlation from the literature. In addition, it is less expensive to define a correlation resulting from numerical calculations rather than from tests. First, the Bi-CITSGM method is validated on the REV, and then, the implementation on CFD simulations of the Dowlati's experiment is discussed.

5.1. Validation of the Bi-CITSGM Method on the REV

Figure 3 depicts the 2-D REV of the tube bundle of Dowlati et al. with symmetric and periodic boundary conditions. The dimensions of transverse pitches and the tube diameter are given in Table 1. CFD calculations were done with OpenFOAM. The governing equations were unsteady and dimensionless, and the k-ω SST turbulence model is used. The flow is computed until the fully developed flow is established. The density of the fluid was always equal to 1 kg/m^3, the desired mass flow was 4.7625×10^{-2} kg/s, and the viscosity varied according to the desired Reynolds number.

All the considered training Reynolds numbers are {2000; 3000; ...; 29,000; 30,000}. This range of training Reynolds numbers fits with the Reynolds number in each cell of the Dowlati's simulation. For each Reynolds number, a thousand time steps are kept, and these snapshots are once and for all decomposed by POD. For each new Reynolds number, the Bi-CITSGM method gives the results almost instantly. In Figures 13–15, solving the ROM with five or ten POD modes are compared to the reference CFD calculation for each Reynolds number. The Bi-CITSGM method enables determining the pressure field with an accuracy less than 10%, except for Reynolds numbers less than 5000 (Figure 13). To increase the accuracy of pressure fields, the spacing between the training Reynolds numbers should be reduced in this range. However, the key parameter in this section is the pressure drop. Figures 14 and 15 show that the pressure drop prediction is much less than 10% even for Reynolds numbers less than 5000. For instance, at a Reynolds number of 24,500, the absolute error between the CFD reference pressure field and the interpolated pressure field at two given times is represented in Figure 16. The highest errors are rather local in the REV, and overall, the interpolated pressure field is very close to the CFD reference. Following the good results achieved by the Bi-CITSGM method on the REV, the results in the next subsection are plotted by keeping only five POD modes.

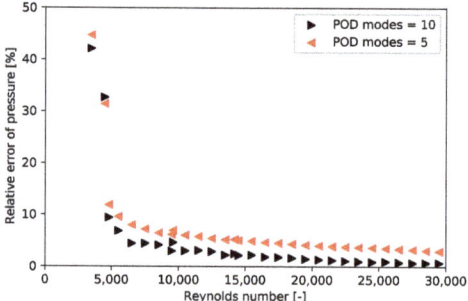

Figure 13. Relative error of the time- and area-weighted pressure average between Bi-CITSGM method and CFD calculation.

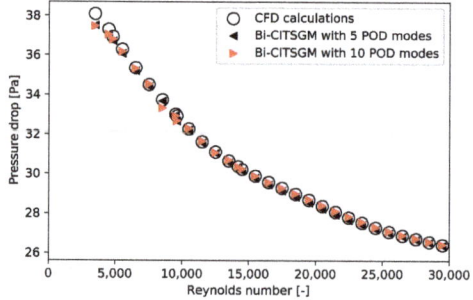

Figure 14. Time-weighted pressure drop average for Bi-CITSGM method and CFD calculations.

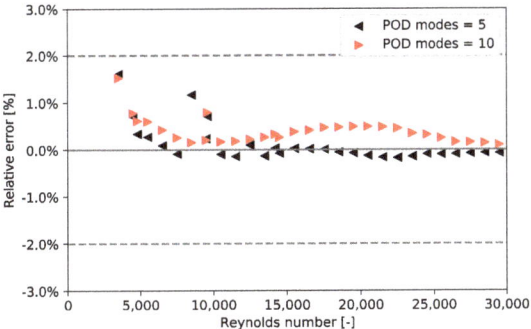

Figure 15. Relative error of the time weighted-pressure drop average between Bi-CITSGM method and CFD calculation.

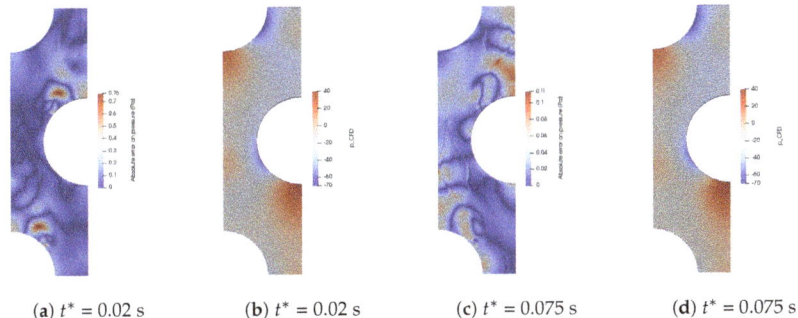

(**a**) $t^* = 0.02$ s (**b**) $t^* = 0.02$ s (**c**) $t^* = 0.075$ s (**d**) $t^* = 0.075$ s

Figure 16. Absolute error between CFD calculations and Bi-CITSGM method for $Re = 24,500$.

5.2. Validation of the Implementation of the Bi-CITSGM Method on the Dowlati's Experiment

The Bi-CITSGM method is solved at each REV of the tube bundle and each iteration in order to compute the Forchheimer term. In addition, the void fraction model of Feenstra et al. is always implemented. Figure 17 compares the total two-phase pressure drop of Dowlati's tube bundle given by the experiment, the Zukauskas correlation implementation and the Bi-CITSGM method implementation in the Forchheimer term. Total pressure drops well fit with the experiment results at low void fractions. Errors become higher when the void fraction increases; however, the post-processing of the pressure drop from the paper of Dowlati et al. [26] is not immediate, nor is it very accurate. The post-process of the pressure drops resulting from the implementation of the Zukauskas correlation and that of the Bi-CITSGM method in the source term are superimposed on the graph. These results validate the suggested approach. Likewise, in Figure 18, the two-phase frictional pressure drops are plotted for different mass flux and are compared to the experiment results. We note that the two-phase frictional pressure drop is highly dependent on the mass flux and less on the void fraction. On the contrary, the gravitational pressure drop decreases when the void fraction increases and is barely dependent on the mass flux. These results are consistent with the physical phenomena that occur in a two-phase flow through a tube bundle. Consequently, the aim to determine the momentum sink by the non-intrusive reduced-order model, Bi-CITSGM, is validated by the results presented in this subsection.

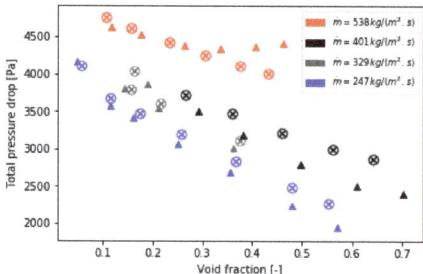

Figure 17. Total pressure drop of Dowlati's tube bundle (×: *Zukauskas*; ○: *Bi-CITSGM*; ▲: *experiment*).

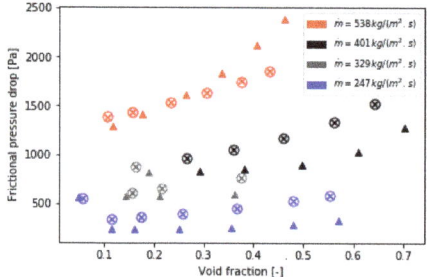

Figure 18. Frictional pressure drop of Dowlati's tube bundle (×: *Zukauskas*; ○: *Bi-CITSGM*; ▲: *experiment*).

6. Conclusions

In order to predict the thermal–hydraulic performance of an adiabatic upward air–water flow through a horizontal tube bundle, two approaches are suggested. They are validated with the experimental results of Dowlati et al. First, the prediction of the void fraction on the tube bundle was improved by using the mixture model and rewriting the drift velocity as a function of slip. Two correlations coming from the literature, Hibiki et al. [17] and Feenstra et al. [16], are compared to the experimental results. They are given similar results with a relative error about 20%. Moreover, we showed that the definition of the Capillary number with the upstream mass flux in Feenstra's correlation significantly improves the void fraction prediction with a relative error under 10%. Second, the CFD porous media approach used implies adding a momentum sink to the governing momentum equation named the Darcy–Forchheimer term. Usually, pressure drop correlations coming from the literature have been used to compute the Forchheimer term except for complex and non-usual geometry for which there is no correlation. In this instance, we demonstrate that it is possible to determine a numerical pressure drop correlation by solving a non-intrusive parametric reduced-order model of the flow through a Representative Elementary Volume of the tube bundle. In the case of the straight tube bundle of Dowlati et al., the Bi-CITSGM method is consistent with the Zukauskas correlation [21]. Moreover, there is a short gap with the experimental results despite a significant possible post-processing error. The two proposed methods that yield satisfactory results need to be expanded. For instance, it would be interesting to simulate a two-phase parallel-flow in a staggered vertical tube bundle with the mixture model modified by the rewriting of the drift velocity. Moreover, the use of a non-intrusive reduced-order model applied to a non-usual geometry of REV in order to compute the Forchheimer term could be an axis of development.

Author Contributions: We would like to declare that all authors have the same contribution listed as follows: C.D., C.A., V.M., C.B. (Claudine Béghein), M.O. and C.B. (Clément Bonneau): Conceptualization; methodology; software; validation; formal analysis; investigation; resources; data curation; writing—original draft preparation; writing—review and editing; visualization; supervision; project administration; funding acquisition. All authors have read and agreed to the published version of the manuscript.

Funding: This research received no external funding.

Institutional Review Board Statement: Not applicable.

Informed Consent Statement: Not applicable.

Data Availability Statement: Not applicable.

Acknowledgments: The authors wish to express their thanks to ANRT (Association Nationale Recherche Technologie) as part of the CIFRE convention N°2018/0443.

Conflicts of Interest: The authors declare no conflict of interest.

Abbreviations

The following abbreviations are used in this manuscript:

CFD	Computational Fluid Dynamic
POD	Proper Orthogonal Decomposition
ROM	Reduced-order model
REV	Representative Elementary Volume
ITSGM	Interpolation on the Tangent Subspace of the Grassmann Manifold
Bi-CITSGM	Bi Calibrated Interpolation on the Tangent Subspace of the Grassmann Manifold
RBF	Radial Basis Function
IDW	Inverse Distance Weighting

Appendix A. Algorithm of the ITSGM Method

Let $\Phi_{\gamma_1}, \Phi_{\gamma_2}, \ldots, \Phi_{\gamma_{N_p}}$ be a set of POD basis and $[\Phi_{\gamma_1}], [\Phi_{\gamma_2}], \ldots, [\Phi_{\gamma_{N_p}}]$ the sub-vector space associated belonging to the Grassmann manifold. By using the definitions of the geodesic path, exponential application, and logarithm application, the aim is to approximate the subspace $[\Phi_{\tilde{\gamma}}]$ amount for a new parameter $\tilde{\gamma} \neq \gamma_i$. The different steps of the ITSGM method come hereafter.

Algorithm A1 ITSGM Algorithm.

(a) Choose a reference point $[\Phi_{\gamma_{i_0}}]$ where $i_0 \in \{1, \ldots, N_p\}$.
(b) For $i \in \{1, \ldots, N_p\}$, determine the vector $\mathcal{X}_i = \mathrm{Log}_{[\Phi_{\gamma_{i_0}}]}(\Phi_{\gamma_i})$ with

$$\mathcal{X}_i = U_i \arctan(\Sigma_i) V_i^T, i = 1, \ldots, N_p \quad (A1)$$

where $U_i \Sigma_i V_i^T$ is the truncated SVD decomposition of $(I - \Phi_{\gamma_{i_0}} \Phi_{\gamma_{i_0}}{}^T) \Phi_{\gamma_i} (\Phi_{\gamma_{i_0}}{}^T \Phi_{\gamma_i})^{-1}$.

(c) Interpolate $\mathcal{X}_1, \mathcal{X}_2, \ldots, \mathcal{X}_{N_p}$ and obtain the initial velocity $\mathcal{X}_{\tilde{\gamma}}$ linked with the new parameter $\tilde{\gamma}$. As the tangent space $\mathcal{T}_{[\Phi_{i_0}]} \mathcal{G}(q, N)$ is a vector space, the interpolation standard technique can be used like Lagrange or RBF.

(d) Determine the interpolate sub-vector space

$$\Phi_{\tilde{\gamma}} = \Phi_{\gamma_{i_0}} \tilde{V} \cos(\tilde{\Sigma}) + \tilde{U} \sin(\tilde{\Sigma}) \quad (A2)$$

where $\tilde{U}\tilde{\Sigma}\tilde{V}^T$ is the SVD decomposition truncated of $\mathcal{X}_{\tilde{\gamma}}$.

Appendix B. Algorithm of the Bi-CITSGM Method

Let the POD decomposition of order q of the matrices S_{γ_i} linked to the parameter γ_i such as

$$S_{\gamma_i} \approx U_{\gamma_i} \Sigma_{\gamma_i} V_{\gamma_i}^T$$

where $U_{\gamma_i} \in \mathbb{R}^{N_x \times q}$ et $V_{\gamma_i} \in \mathbb{R}^{N_s \times q}$ are, respectively, the spatial and temporal bases, and $\Sigma_{\gamma_i} \in \mathbb{R}^{q \times q}$ is the singular value matrix. Obtaining the solution $S_{\widetilde{\gamma}}$ for a new parameter, $\widetilde{\gamma} \neq \gamma_i$ is given by the online step of the Bi-CITSGM method defined below.

Algorithm A2 Bi-CITSGM Algorithm.

Offline step (do this step only once):

(a) For $i = 1, \ldots, N_p$, to approximate the snapshot matrices S_{γ_i}, use the truncated SVD decomposition of order q as follows:

$$S_{\gamma_i} \approx U_{\gamma_i} \Sigma_{\gamma_i} V_{\gamma_i}^T \tag{A3}$$

where $U_{\gamma_i} \in \mathbb{R}^{N_x \times q}$ and $V_{\gamma_i} \in \mathbb{R}^{N_s \times q}$ are, respectively, the spatial and time basis and $\Sigma_{\gamma_i} \in \mathbb{R}^{q \times q}$ is the singular values matrix.

Online step:

(b) Interpolate the singular values matrix Σ_{γ_i} with the classical interpolation method (Lagrange, RBF, etc.) and build the matrix $\widetilde{\Sigma}$ for the new parameter $\widetilde{\gamma}$.

(c) Approximate the spatial basis $U_{\widetilde{\gamma}}$ and the temporal basis $V_{\widetilde{\gamma}}$ with the reduced basis interpolation method ITSGM (described in the Algorithm A1).

(d) Adjust the signs of the modes of the training basis by multiplying the jth spatial mode $\Phi_{\gamma_k}{}^j$ and temporal mode $\Lambda_{\gamma_k}{}^j$ by -1 if the following condition is met:
$||\Phi_{\gamma_{k_0}}{}^j - \Phi_{\gamma_k}{}^j||_2 > ||\Phi_{\gamma_{k_0}}{}^j + \Phi_{\gamma_k}{}^j||_2$ with $k_0 = \underset{i \in \{1,\ldots,N_p\}}{\mathrm{argmin}}\ dist_G(\Phi_{\widetilde{\gamma}}, \Phi_{\gamma_i})$.

(e) Calculate the considered coefficients ω_i and κ_i

$$\omega_i = \frac{dist_G(U_{\widetilde{\gamma}}, U_{\gamma_i})^{-m}}{\sum_{k=1}^{N_p} dist_G(U_{\widetilde{\gamma}}, U_{\gamma_k})^{-m}} \qquad \kappa_i = \frac{dist_G(V_{\widetilde{\gamma}}, V_{\gamma_i})^{-l}}{\sum_{k=1}^{N_p} dist_G(V_{\widetilde{\gamma}}, V_{\gamma_k})^{-l}} \tag{A4}$$

(f) Calculate λ the diagonal matrix of eigenvalues, and P the matrix of eigenvectors by verifying the following eigenvalue decomposition:

$$\sum_{i=0}^{N_p} \sum_{j=0}^{N_p} \omega_i \omega_j U_{\gamma_i}^T U_{\widetilde{\gamma}} U_{\widetilde{\gamma}}^T U_{\gamma_j} = P \lambda P^T \tag{A5}$$

(g) Calculate η the diagonal matrix of eigenvalues, and H the matrix of eigenvectors by verifying the following eigenvalue decomposition:

$$\sum_{i=0}^{N_p} \sum_{i=0}^{N_p} \kappa_i \kappa_j V_{\gamma_i}^T V_{\widetilde{\gamma}} V_{\widetilde{\gamma}}^T V_{\gamma_j} = H \eta H^T \tag{A6}$$

(h) Calculate the orthogonal matrices K and Q given by:

$$K = U_{\widetilde{\gamma}}^T \left(\sum_{i=0}^{N_p} \omega_i U_{\gamma_i} \right) P \lambda^{-\frac{1}{2}} P^T \tag{A7}$$

$$Q = V_{\widetilde{\gamma}}^T \left(\sum_{i=0}^{N_p} \kappa_i V_{\gamma_i} \right) H \eta^{-\frac{1}{2}} H^T \tag{A8}$$

(i) Build the interpolate snapshot matrix defined by:

$$S_{\widetilde{\gamma}} = U_{\widetilde{\gamma}} K \widetilde{\Sigma} Q^T V_{\widetilde{\gamma}}^T \tag{A9}$$

References

1. Carver, M.B.; Carlucci, L.N.; Inch, W.W.R. *Thermal-Hydraulics in Recirculating Steam Generators—THIRST Code User's Manual*; Atomic Energy of Canada Limited: Chalk River, ON, Canada, 1981.
2. Tincq, D.; David, F. THYC, un code 3D de thermohydraulique pour les générateurs de vapeur, les échangeurs de chaleur et les condenseurs. *Rev. Générale De Therm.* **1995**, *34*, 141–153.
3. Zuber, N.; Findlay, J.A. Average Volumetric Concentration in Two-Phase Flow Systems. *J. Heat Transf.* **1965**, *87*, 453–468. [CrossRef]
4. Ozaki, T.; Suzuki, R.; Mashiko, H.; Hibiki, T. Development of drift flux model based on 8x8 BWR rod bundle geometry experiments under prototypic temperature and pressure conditions. *J. Nucl. Sci. Technol.* **2013**, *50*, 563–580. [CrossRef]
5. Lellouche, G.S.; Zolotar, B.A. *Mechanistic Model for Predicting Two-Phase Void Fraction for Water in Vertical Tubes, Channels, and Rod Bundles*; No. EPRI-NP-2246-SR; Electric Power Research Inst.: Palo Alto, CA, USA, 1982.
6. Armand, A.A. The resistance during the movement of a two-phase system in horizontal pipes. *At. Energy Res. Establ.* **1959**, *1*, 16–23.
7. Massena, W.A. *Steam-Water Pressure Drop and Critical Discharge Flow—A Digital Computer Program*; No. HW-65706; General Electric Co. Hanford Atomic Products Operation: Richland, WA, USA, 1960.
8. Mao, K.; Hibiki, T. Drift-Flux Model Upward Two-Phase Cross-Flow Horiz. Tube Bundles. *Int. J. Multiph. Flow* **2017**, *91*, 170–183. [CrossRef]
9. Stosic, Z.V.; Stevanovic, V.D. Advanced Three-Dimensional Two-Fluid Porous Media Method for Transient Two-Phase Flow Thermal-Hydraulics in Complex Geometries. *Numer. Heat Transf. B.* **2002**, *41*, 263–289. [CrossRef]
10. Ishii, M.; Zuber, N. Drag coefficient and relative velocity in bubbly, droplet or particulate flows. *AIChE J.* **1979**, *25*, 843–855. [CrossRef]
11. Schiller, L.; Naumann, A. Fundamental calculations in gravitational processing. *Z. Des Vereines Dtsch. Ingenieure* **1933**, *77*, 318–320.
12. Tomiyama, A.; Kataoka, I.; Zun, I.; Sakaguchi, T. Drag Coefficients of Single Bubbles under Normal and Micro Gravity Conditions. *JSME Int. J. Ser. B* **1998**, *41*, 472–479. [CrossRef]
13. Morsi, S.; Alexander, A.J. An investigation of particle trajectories in two-phase flow systems. *J. Fluid Mech.* **1972**, *55*, 193–208. [CrossRef]
14. Ishii, M.; Hibiki, T. *Thermo-Fluid Dynamics of Two-Phase Flow*, 2nd ed.; Springer: New York, NY, USA, 2011.
15. Ozaki, T.; Hibiki, T.; Miwa, S.; Mori, M. Code performance with improved two-group interfacial area concentration correlation for one-dimensional forced convective two-phase flow simulation. *J. Nucl. Sci. Technol.* **2018**, *55*, 911–930. [CrossRef]
16. Feenstra, P.; Weaver, D.; Judd, R. An improved void fraction model for two-phase cross-flow in horizontal tube bundles. *Int. J. Multiph. Flow* **2000**, *26*, 1851–1857. [CrossRef]
17. Hibiki, T.; Mao, K.; Ozaki, T. Development of void fraction-quality correlation for two-phase flow in horizontal and vertical tube bundles. *Prog. Nucl. Energy* **2017**, *97*, 38–52. [CrossRef]
18. Smith, S.L. Void fractions in two-phase flow: A correlation based upon an equal velocity head model. *Proc. Inst. Mech. Eng.* **1969**, *184*, 647–664. [CrossRef]
19. Darcy, H. *Les Fontaines Publiques de la ville de Dijon*; Victor Dalmont: Paris, France, 1856.
20. Forchheimer, P. Wasserbewegung durch Boden. *Z. Ver. Dtsch. Ing.* **1901**, *45*, 1782–1788.
21. Zukauskas, A.; Ulinskas, R. Banks of plain and finned tubes. In *Heat Exchanger Design Handbook*; Schlunder, E.U., Ed.; Hemisphere Publishing Corporation: Irvine, CA, USA, 1983; pp. 1–17.
22. Rhodes, D.B.; Carlucci, L.N. Predicted and measured velocity distribution in a model heat exchanger. In Proceedings of the International Conference on Numerical Methods in Nuclear Engineering, Montreal, QC, Canada, 6–9 September 1983; AECL-8271.
23. Oulghelou, M.; Allery, C. Non-intrusive method for parametric model order reduction using a bi-calibrated interpolation on the grassmann manifold. *J. Comput. Phys* **2021**, *426*, 109924. [CrossRef]
24. Oulghelou, M.; Allery, C. Hyper bi-calibrated interpolation on the Grassmann manifold for near real time flow control using genetic algorithm. *arXiv* **2019**, arXiv:1901.03177.
25. Amsallem, D.; Farhat, C. Interpolation Method for Adapting Reduced-Order Models and Application to Aeroelasticity. *AIAA J.* **2008**, *46*, 1803–1813. [CrossRef]
26. Dowlati, R.; Chan, A.M.C.; Kawaji, M. Hydrodynamics of two-phase flow across horizontal in-line and staggered rod bundles. *J. Fluids Eng.* **1992**, *114*, 450–456. [CrossRef]
27. Manninen, M.; Taivassalo, V. On the mixture model for multiphase flow. *VTT Publ.* **1996**, *288*, 1–67.
28. Woldesemayat, M.A.; Ghajar, A.J. Comparison of void fraction correlations for different flow patterns in horizontal and upward inclined pipes. *Int. J. Multiph. Flow* **2007**, *33*, 347–370. [CrossRef]
29. Godbole, P.V.; Tang, C.C.; Ghajar, A.J. Comparison of Void Fraction Correlations for Different Flow Patterns in Upward Vertical Two-Phase Flow. *Heat Transf. Eng.* **2011**, *32*, 843–860. [CrossRef]
30. Consolini, L.; Robinson, D.; Thome, J.R. Void fraction and two-phase pressure drops for evaporating flow over horizontal tube bundles. *Heat Transf. Eng.* **2006**, *27*, 5–21. [CrossRef]
31. Ishihara, K.; Palen, J.W.; Taborek, J. Critical review of correlations for predicting two-phase flow pressure drop across tube banks. *Heat Transf. Eng.* **1980**, *1*, 23–32. [CrossRef]

32. Lockhart, R.W.; Martinelli, R.C. Proposed correlation of data for isothermal two-phase, two-component flow in pipes. *Chem. Eng. Prog.* **1949**, *45*, 39–48.
33. Abu-Hamdeh, H.N.; Bantan, R.A.R.; Alimoradi, A. Thermal and hydraulic performance of the twisted tube bank as a new arrangement and its comparison with other common arrangements. *Chem. Eng. Res. Des.* **2020**, *157*, 46–57. [CrossRef]
34. Wu, C.C.; Chen, C.K.; Yang, Y.T.; Huang, K.H. Numerical simulation of turbulent flow forced convection in a twisted elliptical tube. *Int. J. Therm. Sci.* **2018**, *132*, 199–208. [CrossRef]
35. Tan, X.; Zhu, D.S.; Zhou, G.Y. Heat transfer and pressure drop performance of twisted oval tube heat exchanger. *Appl. Therm. Eng.* **2013**, *50*, 374–383. [CrossRef]
36. Patankar, S.V.; Liu, C.H.; Sparrow, E.M. Fully Developed Flow and Heat Transfer in Ducts Having Streamwise-Periodic Variations of Cross-Sectional Area. *J. Heat Transf.* **1977**, *99*, 180–186. [CrossRef]
37. Stalio, E.; Piller, M. Direct Numerical Simulation of Heat Transfer in Converging-Diverging Wavy Channels. *J. Heat Transf.* **2007**, *129*, 769–778. [CrossRef]
38. Sahamifar, S.; Kowsary, F.; Mazlaghani, M.H. Generalized optimization of cross-flow staggered tube banks using a subscale model. *Int. Commun. Heat Mass Transf.* **2019**, *105*, 46–57. [CrossRef]
39. Sirovich, L. Turbulence and the dynamics of coherent structures, Part 1: Coherent structures. Part 2: Symmetries and transformations. Part 3: Dynamics and scaling. *Q. J. Mech. Appl. Math.* **1987**, *45*, 561–590.
40. Holmes, P.; Lumley, J.L.; Berkooz, G.; Rowley, C.W. *Turbulence, Coherent Structures, Dynamical Systems and Symmetry*, 2nd ed.; Cambridge University Press: Cambridge, UK, 2012.
41. Amsallem, D.; Cordial, J.; Carlberg, K.; Farhat, C. A method for interpolating on manifolds structural dynamics reduced-order models. *Int. J. Numer. Methods Eng.* **2009**, *80*, 1241–1258. [CrossRef]
42. Mosquera, M.R.; Hamdouni, A.; El Hamidi, A.; Allery, C. POD basis interpolation via Inverse Distance Weighting on Grassmann manifolds. *Discret. Contin. Dyn.-Syst.-S* **2018**, *12*, 1743–1759. [CrossRef]
43. Mosquera, R.; El Hamidi, A.; Hamdouni, A.; Falaize, A. Generalization of the Neville–Aitken interpolation algorithm on Grassmann manifolds: Applications to reduced order model. *Int. J. Numer. Methods Eng.* **2021**, *93*, 2421–2442. [CrossRef]
44. Oulghelou, M.; Allery, C. A fast and robust sub-optimal control approach using reduced order model adaptation techniques. *Appl. Math. Comput.* **2018**, *333*, 416–434. [CrossRef]
45. Oulghelou, M.; Allery, C. Non-intrusive reduced genetic algorithm for near-real time flow optimal control. *Int. J. Numer. Methods Fluids* **2020**, *92*, 1118–1134. [CrossRef]
46. Menter, F.R. Two-Equation Eddy-Viscosity Turbulence Models for Engineering Applications. *AIAA J.* **1994**, *32*, 1598–1605. [CrossRef]
47. Roser, R. Modélisation Thermique de Bouilleurs à Tubes Horizontaux. Etude Numérique et Validation ExpéRimentale. Ph.D. Thesis, Université de Provence Aix-Marseille 1, Marseille, France, 1999.

Article

Analysis of Energy Dissipation of Interval-Pooled Stepped Spillways

Xin Ma [1,2], Jianmin Zhang [1,*] and Yaan Hu [2,*]

1. State Key Laboratory of Hydraulics and Mountain River Engineering, Sichuan University, Chengdu 610065, China; xma@nhri.cn
2. State Key Laboratory of Hydrology-Water Resources and Hydraulic Engineering, Nanjing Hydraulic Research Institute, Nanjing 210029, China
* Correspondence: zhangjianmin@scu.edu.cn (J.Z.); yahu@nhri.cn (Y.H.)

Abstract: The water flow characteristics over an interval-pooled stepped spillway are investigated by combining the renormalization group (RNG) k-ε turbulence model with the volume of fluid (VOF) interface capture technique in the present study. The results show that the energy dissipation performance of the interval-pooled stepped spillway was generally better than that of the pooled, stepped spillways and the traditional flat-panel stepped spillway. The omega vortex intensity identification method is introduced to evaluate the energy dissipation. Due to the formation of "pseudo-weir", the energy dissipation did not increase with the growth of the pool's height. In addition, the average vortex intensity can characterize the dissipation rate to some extent.

Keywords: energy dissipation; interval-pooled stepped spillway; numerical simulation; omega identification method

Citation: Ma, X.; Zhang, J.; Hu, Y. Analysis of Energy Dissipation of Interval-Pooled Stepped Spillways. *Entropy* **2022**, *24*, 85. https://doi.org/10.3390/e24010085

Academic Editors: Amine Ammar, Francisco Chinesta and Rudy Valette

Received: 29 November 2021
Accepted: 28 December 2021
Published: 4 January 2022

Publisher's Note: MDPI stays neutral with regard to jurisdictional claims in published maps and institutional affiliations.

Copyright: © 2022 by the authors. Licensee MDPI, Basel, Switzerland. This article is an open access article distributed under the terms and conditions of the Creative Commons Attribution (CC BY) license (https://creativecommons.org/licenses/by/4.0/).

1. Introduction

Stepped spillways have been widely utilized as energy dissipation facilities in hydraulic engineering and show great potential due to achieving a better rate of energy dissipation while releasing excess flood water [1–3]. They can reduce the scale of the stilling basin and the number of downstream protection works and decrease the extent of downstream river erosion, which has excellent economic and technical performance indicators [4]. To improve the energy dissipation effect and hydraulic characteristics of stepped spillways, several studies have been conducted to optimize the configurations. The stepped spillway is not confined to flat, uniform steps, and some models of stepped chutes have been designed with changing channel slopes [5], nonuniform steps [6], and pooled steps [7,8]. Among them, Felder and Chanson (2013) and Thorwarth (2009) conducted physical experiments on pooled stepped spillways with chute slopes of θ = 8.9°, 14.6°, and 26.8°, and the results showed that the energy dissipation efficiency of pooled stepped spillways performed well, but unstable free surface fluctuations occurred at a chute slope θ of 8.9°, which could cause some potential problems for the step structure. Moreover, a new type of pooled stepped spillway which has a pool on the horizontal step face of every second step was also proposed, and its flow characteristics were investigated by several researchers [9–11]. Several experimental studies have been conducted to analyze the flow pattern of pooled stepped spillways, and the corresponding step spillway parameters in these studies are presented in Table 1. Among them, Kökpinar (2004) made a detailed comparison of the air-liquid flow parameters for a 30° interval-stepped spillway with a classical step stepped spillway and a pooled stepped spillway, and the results indicate that the interval enclosure spillway can entrain more air and reduce the risk of cavitation. André and Schleiss (2004) then conducted physical experiments for interval-stepped spillways with θ = 30° and θ = 18.6°, and they found that interval-stepped spillways have better energy dissipation performance with a defined pool height. Felder and Chanson (2013) conducted experiments on an interval-stepped spillway with a small angle ($\theta = 8.9°$). Their cases were tested in nappe and transition flow regimes and did not achieve a skimming flow due to the chute angle and discharge. All the above-mentioned studies realized that interval-stepped spillways have totally different hydraulic characteristics from pooled stepped spillways and flat stepped spillways, their energy dissipation aspect is still lacking, and systematic research is needed.

Table 1. Summary of experimental studies of flow properties on interval-pooled stepped spillway configurations.

Reference	θ (°)	Step Geometry	Comment	Flow Regime	Methodology
Kökpinar (2004)	30	$h = 6$ cm, $l = 10.4$ cm, $d = 3$ cm	$W = 0.5$ m, 64 steps, $w_p = 2.6$ cm	NA/TRA/SK	physical model experiment
André and Schleiss (2004)	18.6/30	$h = 6$ cm, $l = 17.8$ cm, $d = 3$ cm, $h = 6$ cm, $l = 10.4$ cm, $d = 3$ cm	$W = 0.5$ m, 42/64 steps, $w_p = 2.6$ cm	NA/TRA/SK	physical model experiment
Felder and Chanson (2013)	8.9	$h = 5$ cm, $d = 5$ cm, $l = 31.9$ cm	$W = 0.5$ m, 21 steps, $w_p = 1.5$ cm	NA/TRA	physical model experiment

Notes: θ = angle between pseudo-bottom formed by the step edges and the horizontal; h = vertical step height (m); l = horizontal step length (m); d = pool height (m); w_p = width of the pool weir crest (m); W = channel width (m); SK = skimming flow regime; TRA = transition flow regime; NA = nappe flow regime.

On the other hand, after decades of development, numerical simulation has become another important tool for the study of hydraulic phenomena [12,13]. During this period, the flow and air entrainment over stepped spillways have been studied by numerical simulations using different methods, such as Reynolds-averaged Navier–Stokes (RANS) [14,15] and meshless smooth particle hydrodynamics (SPH) [16]. This not only shows more clearly the development pattern within the flow field but also the interaction of the hydraulic conditions. A large number of vortex structures with different scales exist in the skimming flow, which plays a key role in the energy dissipation efficiency. Therefore, accurate identification of the vortex intensity is of great importance for understanding the flow mechanisms and analyzing spillway dissipation problems. Vortex identification techniques have also undergone rapid development in recent decades such as the vorticity threshold method, the Q criterion method, the λ criterion method, the Ω criterion method, and the Rortex method [17–20], which analogizes the above mainstream vortex identification methods, among which the Ω criterion method has better performance in capturing the vortex structure generated near the NACA66(mod) edge of the water wing at high speed and identifying the strong rigid body rotation and weak rotation regions. Therefore, this paper mainly uses the Ω criterion method to explore the step vortex structure [21].

To the best of our knowledge, the influence of various pool heights on the energy dissipation of interval-stepped spillways is still unknown. Thus, the vortex intensity and distribution are explored in the present study through numerical methods. Furthermore, the relationship between the pool height and energy dissipation rate is also investigated preliminarily.

2. Numerical Simulation
2.1. Computational Domain

Due to their high performance and effectiveness, numerical methods are commonly used in hydraulic and hydrological studies [22–24]. Consequently, numerical simulations are adopted in this paper to examine the air-water two-phase flows over the spillway. The computation domain is described in detail and depicted in Figure 1. The experimental data were applied to validate the results from a numerical simulation [25]. Therefore, the shape parameters of the interval-pooled stepped spillways (width, length of the step, and height of the step and chute slopes) are consistent with the referred experimental study. Moreover, four pool heights (d = 0.025, 0.05, 0.075 and 0.1 m) are applied to investigate its influence on the flow pattern (Figure 1). As indicated by an earlier study [26,27], an equilibrium state will be reached between the water head loss and gravity if the step spillway is long enough. After the validation of the reliability of the numerical simulation, 10 steps were added downstream of the original 10 steps to ensure that the water flow near the downstream area formed a "quasi-uniform flow". The computational domain of this model is shown in Figure 1, with an upstream tank with a volume of 0.58 × 0.52 × 0.50 m³ and a spillway crest with a length of 1.01 m followed by 20 steps, where each step is 0.2 m long (l) and 0.1 m high (h). The overall spillway slope (θ) is 26.6°, and the width (W) is 0.52 m. The width of the pool weir crest (w_p) is 0.015 m. The channel width (W), chute slope h/l, and pool weir crest w_p were kept the same in different shapes and cases (Figure 1).

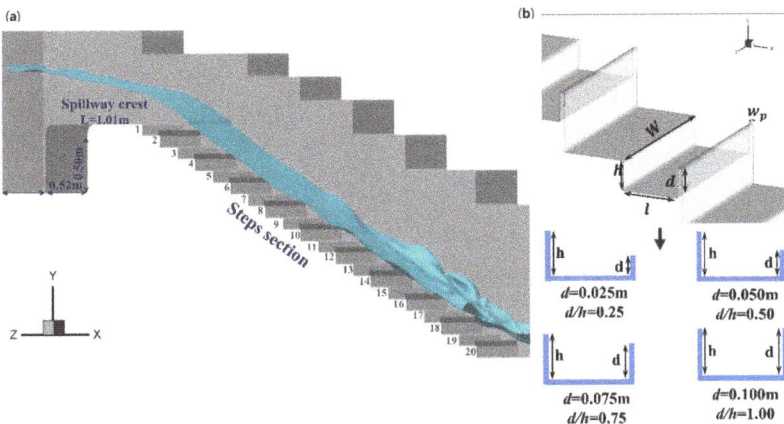

Figure 1. (a) Schematic of an interval-pooled stepped spillway. (b) Sketch of the specific step size.

2.2. Boundary Conditions

In the present study, the dimensionless depth d_c/h is used, where h is the step height and d_c is the critical water depth. The critical water depth can be calculated by $d_c = \sqrt[3]{Q^2/(g \times W^2)}$, where Q is the water discharge and W is the step width. Therefore, $d_c/h = \left(q/\sqrt{g \times h^3}\right)^{2/3}$ and is proportional to the Froude number. Our study only addresses the skimming flow, since it aims to investigate the hydraulic energy dissipation efficiency and hydrodynamic characteristics with a high flow rate with a discharge range of $1.90 \leq d_c/h \leq 5.14$. The specific values that are given are shown in Table 2. Similarly, these parameters may be indicative of a full scale to guarantee that scale effects are unlikely to influence the extrapolation of the results to prototype conditions (Felder, Guenther, et al., 2012). In terms of model selection, the volume of fluid (VOF) method [28] was chosen for free surface tracking. Since the RNG model added an additional term to the equation, it allowed the whole model to simulate more complex flows more accurately [29]. In many studies, the results simulated using the RNG model have produced reliable results [30,31]. Therefore the renormalization group (RNG) k-ε turbulence model [32] was applied. The upper inlet boundary was set as the velocity inlet boundary, while the no-slip boundary and standard wall function for the sticky bottom layer were adopted on the wall. In addition, the pressure outlet boundary condition was employed for the outlet boundary, and the pressure inlet boundary was applied for the air-inlet boundary (Figure 2).

Table 2. The flow conditions for different channel configurations.

Q (m³/s)	d_c/h	Flat	Pooled	d = 0.25 h	d = 0.50 h	d = 0.75 h	d = 1.00 h
Q_1 = 0.123	1.79			✓	✓	✓	
Q_2 = 0.135	1.90	✓	✓	✓	✓	✓	✓
Q_3 = 0.148	2.02			✓	✓	✓	✓
Q_4 = 0.160	2.13	✓	✓	✓	✓	✓	✓
Q_5 = 0.188	2.37			✓	✓		
Q_6 = 0.216	2.60			✓	✓		
Q_7 = 0.244	2.82	✓	✓	✓	✓		✓
Q_8 = 0.272	3.03			✓	✓	✓	
Q_9 = 0.300	3.24	✓	✓	✓	✓	✓	✓

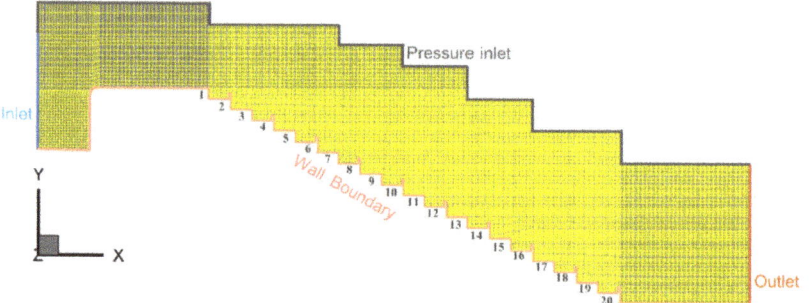

Figure 2. Meshing pattern in the computational domain.

2.3. Mesh and Model Validation

For the simulation of the flow over pooled, stepped spillways, a nonuniform structured mesh was applied. Since the mesh density has a significant effect on the accuracy and reliability of the results, mesh independence was applied in this section. The grid convergence index (GCI) method, based on the Richardson extrapolation (RE) method, is an appropriate and recommended method that has been evaluated over several hundred computational fluid dynamics (CFD) cases [33,34]. The GCI formula is described as follows:

$$GCI = \frac{1.25 \left| \frac{\varnothing_1 - \varnothing_2}{\varnothing_1} \right|}{\left(\frac{h_2}{h_1} \right)^{p'} - 1} \tag{1}$$

where $p' = \frac{1}{\ln(h_2/h_1)} \ln \left| \frac{\varnothing_3 - \varnothing_2}{\varnothing_2 - \varnothing_1} \right| + \ln \left(\frac{\left(\frac{h_2}{h_1}\right)^p - sgn\left(\frac{\varnothing_3 - \varnothing_2}{\varnothing_2 - \varnothing_1}\right)}{\left(\frac{h_3}{h_2}\right)^p - sgn\left(\frac{\varnothing_3 - \varnothing_2}{\varnothing_2 - \varnothing_1}\right)} \right)$, $\varnothing_i$ is the calculation result of the i-th grid, i is taken as 1, 2, and 3, and h_i is the average grid size of the i-th grid and satisfies the relationship of $h_1 < h_2 < h_3$.

To check whether the numerical results were influenced by the grid density, three different sizes of structured grids were tested. The average cell grid sizes were 0.0186 m, 0.0152 m, and 0.012 m, and the corresponding total numbers of elements were 280,000, 480,000 (as shown in Figure 2), and 680,000, respectively. According to the GCI method, the maximum GCI values in the velocity profiles were 6.7 and 3.5% (Figure 3). Considering the efficiency and accuracy of the simulation, the average cell grid of 0.0152 m was used for all subsequent analyses.

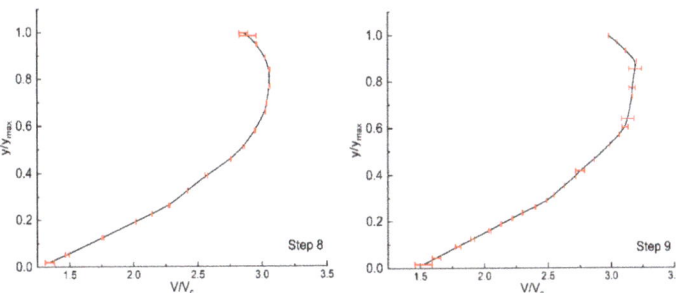

Figure 3. Discretization error bars computed using the GCI index on the 8th and 9th step horizontal plane.

The simulation was validated against the experimental data of Felder, Guenther et al. (2012) in terms of both the velocity profile and energy dissipation. Figure 4 shows that the simulation results agreed well with the experimental data, and the maximum error of the flow velocity on the 8th step and 9th step was only 7.29% and 6.58%, respectively. The energy dissipation rates of the flat stepped spillway (d_c/h = 0.81) and pooled, stepped spillway (d_c/h = 1.85) were calculated according to Equation (2). The errors were 9.17% and 8.62%, respectively, which were within a reasonable range.

$$H_{res} = d_d \times \cos\theta + \frac{U_w}{2 \times g} + d \tag{2}$$

where θ represents the angle, U_w represents the velocity (m/s)—that is, the average velocity of the vertical distance from the edge of the step to 90% of the mainstream water depth—d is the pool height, d_d is the water flow depth, and H_{res} is the residual head (m).

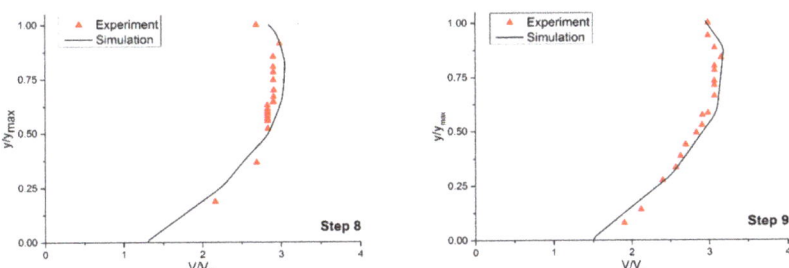

Figure 4. Experimental and numerical simulation comparison chart of the velocity on the 8th and 9th step horizontal plane.

3. Results and Discussion

3.1. Energy Dissipation Performance

In Figure 5a, the vertical coordinate in Figure 5a represents dimensionless residual energy H_{res}/d_c, and the horizontal coordinate is d_c/h. The present data show there was a large difference in the residual energy of the connected steps of the interval-pooled stepped spillways. This means that the method is not applicable to measure the energy dissipation rate in interval-pooled stepped spillways. The experimental results obtained by Thorwarth (2009) are also presented in the same diagram. The residual water head of the pool case decreased with the increasing flow, leading to a higher dissipation rate. It is noteworthy that the residual head of the pooled, stepped spillways increased with the height of the pool even at other angles (θ = 14.6°); that is, in pooled, stepped spillways, the increase in the pool height did not result in a significant increase in dissipation. The reason for this can be attributed to the fact that with the increase in the pool height, the step cavity circulation became stable, and the dissipation energy in the mainstream area was greater than the circulation in the step cavity (Thorwarth 2009). The mechanism of energy dissipation in the interval steps will be further discussed later.

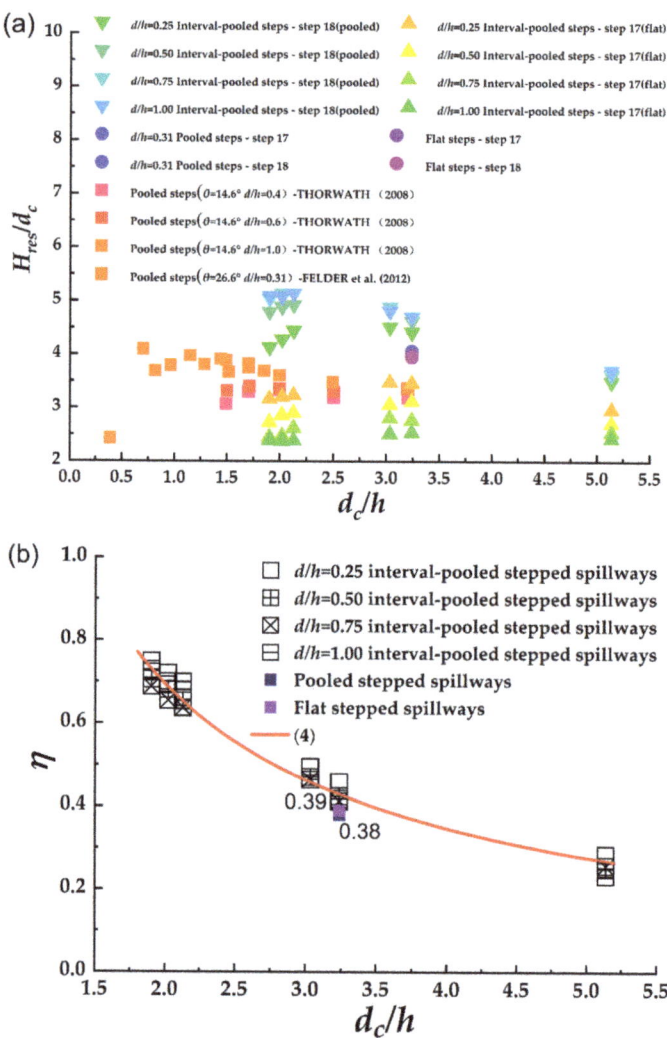

Figure 5. (a) Dimensionless residual energy of stepped spillways. (b) Energy dissipation rate of interval-pooled stepped spillways.

The above-mentioned approach could obtain the energy dissipation rate of a specific step but could not evaluate the performance of the whole stepped spillway. From Figure 5a, it can be found that the residual head of the 17th step and the 18th step appeared to contrast greatly. Clearly, it was more difficult to evaluate the dissipative performance of interval-pooled stepped spillways using a certain independent step-residual head. Therefore, it was necessary to estimate the energy dissipation of the whole spillway (i.e., to calculate the energy difference between the water flow upstream and downstream entering the stepped spillway (as shown in Figure 6)), and the energy dissipation rate of the whole spillway could be calculated according to Equation (3), converted through the Bernoulli equation:

$$\eta = \frac{\Delta E}{E_1} \times 100\% = \frac{E_1 - E_2}{E_1} \times 100\% \tag{3}$$

where η is the energy dissipation rate, which means the percentage difference between the pre-step and post-step energy and the total pre-step energy, ΔE is the energy difference between the pre- and post-stepped spillways, and E_1 and E_2 are the total energy upstream and downstream of the stepped spillways, respectively:

$$\eta = 1.39\left(\frac{d_c}{h}\right)^{-1} \tag{4}$$

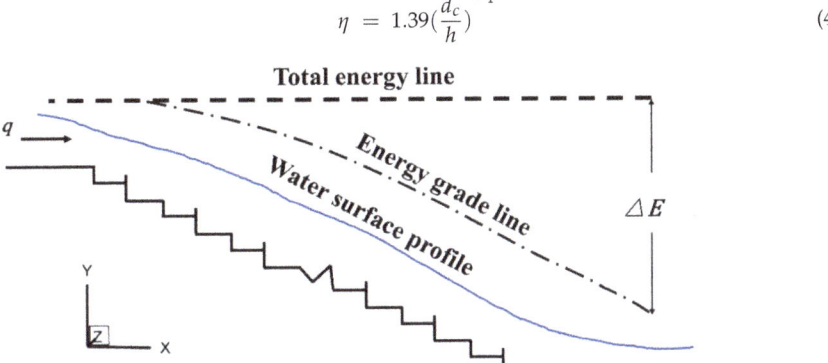

Figure 6. Overall interval-pooled stepped spillways energy dissipation rate calculation schematic.

The energy dissipation calculated by Equation (3) is shown in Figure 5b. The energy dissipation did not change with the pool height d. In the maximum flow ($d_c/h = 3.24$) scenario, the interval-pooled stepped spillways with $d/h = 0.25$ had the best performance, with an energy dissipation rate of up to 28%. The energy dissipation rate of the interval-pooled stepped spillways was predicted by Equation (4) ($R^2 \sim 0.98$). The energy dissipation rate of the interval-pooled stepped spillway with $d/h = 0.25$ was 20.07% higher than that of the flat stepped spillway, with the same flow rate and step angle. Compared with the pooled, stepped spillway, the energy dissipation rate of the interval-pooled stepped spillway with $d/h = 0.25$ increased by 16.51%. Compared with previous studies, the interval-pooled stepped spillway energy dissipation was only increased by approximately 2% for the step angle of 30.0° in the experiments conducted by André and Schleiss (2004).

3.2. Energy Dissipation Analysis Using the Omega Vortex Identification Method

The dissipation rate of the stepped spillway is closely related to the circulation region of the step (Thorwarth 2009). Therefore, the omega vortex identification method was introduced for stepwise exploration [19]. The guidelines of the Ω method were defined as follows:

$$\Omega = \frac{b}{a + b + \varepsilon} \tag{5}$$

$$a = \text{trace}\left(A^T A\right) = \sum_{i=1}^{3} \sum_{j=1}^{3} \left(A_{ij}\right)^2 \tag{6}$$

$$b = \text{trace}\left(B^T B\right) = \sum_{i=1}^{3} \sum_{j=1}^{3} \left(B_{ij}\right)^2 \tag{7}$$

where A and B are the strain rate and vorticity (or spin) tensors, respectively. The flow is irrotational if all the terms in B are zero. A and B are the symmetric and anti-symmetric parts of the velocity gradient tensor. In addition, a and b are the squares of the Frobenius norm of A and B. ε is a very small number to prevent division by zero. Ω is the ratio of vorticity over the whole motion of the fluid element [19]. Note that $\Omega \in [0,1]$. The flow is a pure deformation when $\Omega = 0$, and the flow is rigidly rotational when $\Omega = 1$. $\Omega > 0.5$ is the region of rigid body rotation, and the more it tends toward 1, the greater the region of pure rigid body rotation:

$$\varepsilon = 10^{-7}(b-a)_{max} \tag{8}$$

It was mentioned by Wang, et al. (2019) that different choices of Ω values have an effect on the vortex structure, and it is appropriate to take the -7th power of the difference between the b and a maxima.

From Figure 7a, for $d_c/h = 3.24$, the vortex structure in the flat stepped spillway had a stable and uniform distribution, which was mainly distributed in the cavity corner and main flow. The vortex was caused by the velocity difference on the edge of the step, which formed a large rotation area. The maximum at the core of the vortex $\Omega_{max} \approx 0.68$. For the pooled stepped spillway ($d/h = 0.31$), the pools increased the strength of the eddies in the cavity corner and decreased the strength of the eddies in the mainstream. $\Omega_{max} \approx 0.72$, which is larger than that of the traditional stepped spillway.

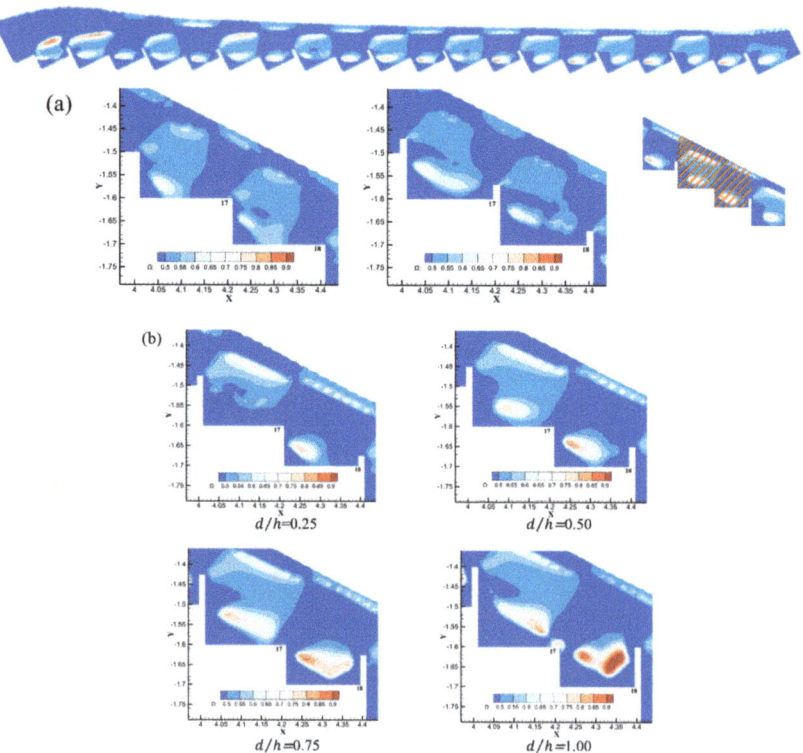

Figure 7. (a) The 17th and 18th step longitudinal interface diagram at the central axis $\Omega > 0.5$ vortex structure in flat stepped spillways and pooled, stepped spillways. The upper inset shows a distribution of $\Omega > 0.5$ vortex intensity in the water flow in the whole area of the step structure ($d/h = 0.5$). The right inset shows a distribution of $\Omega > 0.5$ vortex intensity in the water flow in the 17th and 18th steps ($d/h = 0.5$). (b) Distribution of the 17th and 18th step vortex structure and vortex intensity at the longitudinal median axis for different d/h ratios at $d_c/h = 3.24$.

Then, we refocused on the vortex distribution and strength of the 18th and 19th steps of the interval-pooled stepped spillway in Figure 7b. For $d/h = 0.25$, the vortex corresponding to the 17th step was not in the chamber but in the region of the main flow. The 18th step corresponded to vortices all appearing inside the cavity, with a maximum value Ω_{max} of about 0.76. When $d/h \geq 0.50$, the vortex intensity distribution and the maximum at the core of the vortex appeared within the 18th step with the increase in the 18th pool, but the vortex did not exist within the cavity corner.

When $d/h \geq 0.50$, the vortex intensity distribution started to appear within the 18th step. The maximum at the core of the vortex increased with the increased height of the 19th step pool ($d/h = 0.50$ for $\Omega_{max} \approx 0.82$, $d/h = 0.75$ for $\Omega_{max} \approx 0.86$, and $d/h = 1.00$ for $\Omega_{max} \approx 0.95$) (Table 3). The core of the vortex started to shift outward from the cavity corner with the increase in the height of the 18th step. From the omega maximum distribution (Table 3), the increase in the pool height could effectively enhance the spin-rolling intensity in the cavity. The increased height of the pool blocked more of the water flow over the steps and enhanced the energy dissipation effect of the no-pool steps. The overall energy dissipation rate was improved. This may be the main reason why the interval-pooled stepped spillway's dissipation effect was better than the pooled, stepped spillway and the flat stepped spillway.

Table 3. The omega maximum within the 17th and 18th step cavity.

Flat	Pooled	$d/h = 0.25$	$d/h = 0.50$	$d/h = 0.75$	$d/h = 1.00$
0.68	0.72	0.76	0.82	0.86	0.95

3.3. Formation of a "Pseudo-Weir"

From Figure 8, we can see that the area where the flow velocity was greater than 0 in the vertical upward direction near the edge of the 17 steps increased with the increasing pool height. When $d/h \geq 0.50$, the velocity value in the vertical upward direction was greater than 0, which exceeded the height of the 17 steps, and a "pseudo-weir" was formed at the edge of the 17 steps. Figure 8 shows that as the pool height of the 18 steps increased, the velocity of the water flowing through the 18 steps and then into the 17 steps increased. This phenomenon indicates that the formation of a "pseudo-weir" enhanced the vortex strength in the no-pool steps. This may be the main reason for the shift of the vortex distribution of the 18th step between $d/h = 0.25$ and $d/h = 0.50$.

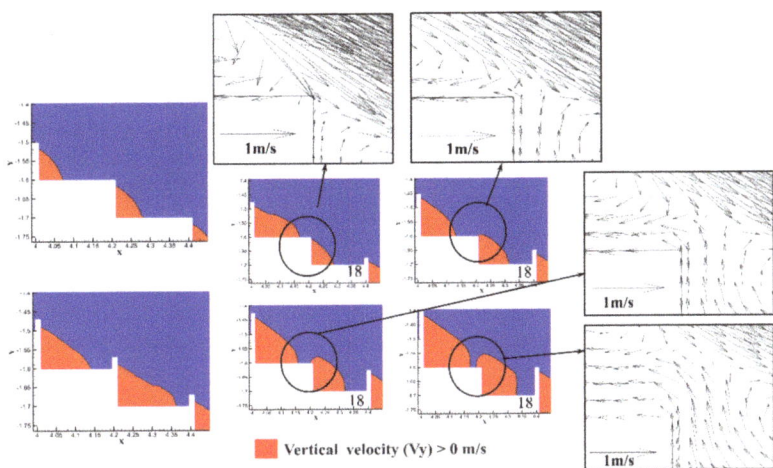

Figure 8. Distribution of vertical velocity Vy at the 18th step. The inset is the distribution of velocity vectors about the 18th step.

As shown in Figures 8 and 9, the increase in pool height caused an increase in the "pseudo-weir". However, the rising "pseudo-weir" induced a continuous increase in stagnant water within all step cavities. Therefore, the increased pool height did not significantly raise the energy dissipation within the step. This was confirmed from Figure 9. It was found that the percentage of negative streamwise velocity at the 17th and 18th steps increased as the pool height increased, which means that the upward shift of the core of vortex position. And the area involved in dissipating energy in the water body would not change significantly with the increase in pool height.

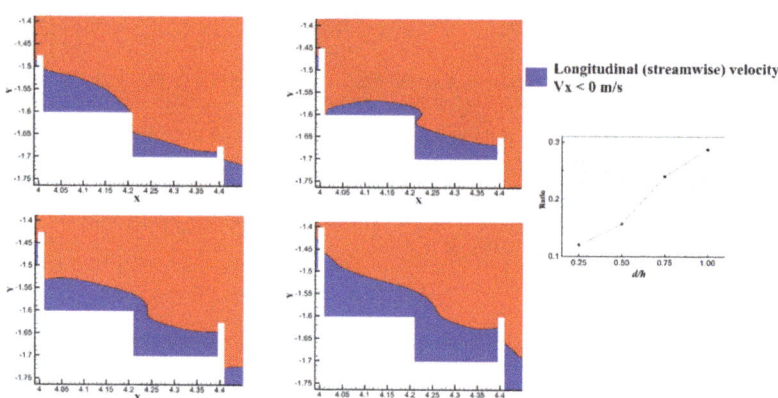

Figure 9. Distribution of streamwise velocity Vx at the 17th and 18th steps. The inset is the percentage of negative velocity vs. the pool height.

3.4. Quantifying Vortex Strength

To better quantify and count the vortex intensity, we created three reference quantities: the area to vortex ratio Ω_p in Equation (9), the average vortex intensity $\Omega_{0.5}$ in the region of $\Omega > 0.50$ in Equation (10), and the average intensity of $\Omega > 0.5$ in the region Ω_d in Equation (11). They are defined as follows:

$$\Omega_p = S_{0.5}/S_c \tag{9}$$

$$\Omega_{0.5} = \frac{1}{N}\sum_{k=1}^{N} \Omega_k \tag{10}$$

$$\Omega_d = \Omega_p \times \Omega_{0.5} \tag{11}$$

where $S_{0.5}$ is the area occupied by $\Omega > 0.5$, S_c is the calculation area, Ω_p denotes the percentage of omega > 0.5 in the calculation region, $\Omega_{0.5}$ is the average of $\Omega > 0.5$, S_k is the kth value of of $\Omega > 0.5$ in the computational domain, and Ω_d means the average intensity of $\Omega > 0.5$ in the region.

The above 3 parameters were calculated for the 17th and 18th steps with whole-step spillways (Tables 4 and 5). The specific positions are shown in Figure 7a with the upper inset and the right inset.

Table 4. Vortex parameters in the water of the 17th and 18th steps for stepped spillways.

	$d/h = 0.25$	$d/h = 0.50$	$d/h = 0.75$	$d/h = 1.00$	Flat ($d/h = 0.00$)	Pooled ($d/h = 0.31$)
Ω_p	0.34	0.44	0.43	0.39	0.49	0.41
$\Omega_{0.5}$	0.58	0.58	0.59	0.60	0.54	0.54
Ω_d	0.20	0.26	0.25	0.24	0.27	0.22
η	0.46	0.42	0.41	0.42	0.39	0.38

Table 5. Vortex parameters in the water of the stepped spillways.

	d/h = 0.25	d/h = 0.50	d/h = 0.75	d/h = 1.00	Flat (d/h = 0.00)	Pooled (d/h = 0.31)
Ω_p	0.38	0.41	0.39	0.36	0.43	0.39
$\Omega_{0.5}$	0.57	0.58	0.59	0.60	0.54	0.55
Ω_d	0.22	0.24	0.23	0.21	0.23	0.22
η	0.46	0.42	0.41	0.42	0.39	0.38

The 17th and 18th steps and the overall step spillway were calculated using Ω_p, $\Omega_{0.5}$, and Ω_d, respectively. It can be seen from Tables 4 and 5 that the values corresponding to the same type of stepped spillway did not differ much. This means that the vortex structure of each of the two steps was more stable and could characterize the development of the overall vortex structure. Ω_p was largest for the flat stepped spillways. Traditional and pooled, stepped spillways accounted for a relatively large number because each set of steps was more uniformly distributed in a better fashion than the vortex distribution range produced by the edge of the steps in addition to the step cavity. In addition to the stepped cavity, the vortex structure with a lower vortex intensity also existed uniformly in the mainstream. However, the $\Omega_{0.5}$ values of conventional stepped spillways and pooled, stepped spillways were small, indicating that the spaced steps strengthened the vortex intensity of all steps. Ω_d is the product of the area ratio and average intensity, which can reflect an average intensity of $\Omega > 0.50$. At the same flow rate, all types of Ω_d did not differ much and belonged to the same order of magnitude. In a comprehensive comparison, the trend of the $\Omega_{0.5}$ index was closer to the change in the energy dissipation rate of each type of stepped spillway, and it could approximate the magnitude of the overall stepped spillway's dissipation rate.

One of our optimization goals was to increase the energy dissipation rate of the spillway. Changing the dimensions to emphasize the effect of the macro-roughness elements on the high-velocity water flow is an important optimized measurement. In further research, we should explore the hydraulic characteristics of more types of step spillways to ensure dam safety and to provide design references.

4. Conclusions

In this study, the step hydraulic characteristics of the traditional step-type spillway were investigated by adding a pool weir to each step. The interval pool-type spillway with different pool heights was investigated by numerical simulation and compared with the conventional flat step and continuous pool-type step. The performances of energy dissipation, vortex distribution, and flow field analysis were discussed in more detail. Based on these studies, the following conclusions can be drawn:

1. The interval-stepped spillway allowed the flow to perform sufficient energy dissipation by longitudinal abrupt expansion and contraction, creating a robust vortex zone in the step cavity. The overall energy dissipation rate had an exponential decay with d_c/h and was generally better than conventional spillways and pooled, stepped spillways.
2. After $d/h \geq 0.50$, each step without a pool formed a "pseudo-weir", which formed a "pseudo-continuous weir" with an increasing pool, increasing the strength of the vortex on the one hand, and on the other hand, the stagnant water body also increased, resulting in an interval-pooled stepped spillway efficiency effect that did not change significantly with the change in pool height.
3. A comprehensive analysis of the step spillway vortex structure was conducted. Three parameters were defined to quantify the variation in vortices within the step, whose $\Omega_{0.5}$ could represent the dissipation rate approximately. This shows that the average intensity of the vortex was closely related to the dissipation effect.

Author Contributions: Funding acquisition, J.Z. and Y.H.; Supervision, J.Z. and Y.H.; Validation, X.M.; Visualization, X.M.; Writing—original draft, X.M.; Writing—review & editing, X.M. All authors have read and agreed to the published version of the manuscript.

Funding: This research was funded by National Science Fund for Distinguished Young Scholars (Grant number No.: 51625901).

Data Availability Statement: Some or all data, models, or code that support the findings of this study are available from the corresponding author upon reasonable request.

Conflicts of Interest: The authors declare no conflict of interest.

References

1. Boes, R.M.; Hager, W.H. Hydraulic Design of Stepped Spillways. *J. Hydraul. Eng.* **2003**, *129*, 671–679. [CrossRef]
2. Chanson, H. Hydraulics of Skimming Flows over Stepped Channels and Spillways. *J. Hydraul. Res.* **1994**, *32*, 445–460. [CrossRef]
3. Chanson, H. *Energy Dissipation in Hydraulic Structures*; CRC Press: Boca Raton, FL, USA, 2015; ISBN 978-1-315-68029-3.
4. Felder, S.; Fromm, C.; Chanson, H. *Air Entrainment and Energy Dissipation on a 8.9 Slope Stepped Spillway with Flat and Pooled Steps*; Department of Civil Engineering, University of Queensland: Brisbane, Australia, 2012; ISBN 978-1-74272-053-1.
5. Chinnarasri, C.; Wongwises, S. Flow Patterns and Energy Dissipation over Various Stepped Chutes. *J. Irrig. Drain. Eng.* **2006**, *132*, 70–76. [CrossRef]
6. Felder, S.; Chanson, H. Energy Dissipation down a Stepped Spillway with Nonuniform Step Heights. *J. Hydraul. Eng.* **2011**, *137*, 1543–1548. [CrossRef]
7. Thorwarth, J. *Hydraulisches Verhalten von Treppengerinnen Mit Eingetieften Stufen: Selbstinduzierte Abflussinstationaritäten Und Energiedissipation*; University of Aachen: Aachen, Germany, 2009.
8. Felder, S.; Chanson, H. Aeration, Flow Instabilities, and Residual Energy on Pooled Stepped Spillways of Embankment Dams. *J. Irrig. Drain. Eng.* **2013**, *139*, 880–887. [CrossRef]
9. André, S.; Schleiss, A. *High Velocity Aerated Flows on Stepped Chutes with Macro-Roughness Elements*; Ecole Polytechnique Federale de Lausanne, Laboratoire de Constructions Hydrauliques: Lausanne, Switzerland, 2004.
10. Kökpinar, M.A. Flow over a Stepped Chute with and without Macro-Roughness Elements. *Can. J. Civ. Eng.* **2004**, *31*, 880–891. [CrossRef]
11. Felder, S.; Chanson, H. Air–Water Flow Measurements in a Flat Slope Pooled Stepped Waterway. *Can. J. Civ. Eng.* **2013**, *40*, 361–372. [CrossRef]
12. Morovati, K.; Eghbalzadeh, A.; Javan, M. Numerical Investigation of the Configuration of the Pools on the Flow Pattern Passing over Pooled Stepped Spillway in Skimming Flow Regime. *Acta Mech.* **2016**, *227*, 353–366. [CrossRef]
13. Gualtieri, C.; Chanson, H. Physical and Numerical Modelling of Air-Water Flows: An Introductory Overview. *Environ. Model. Softw.* **2021**, *143*, 105109. [CrossRef]
14. Meireles, I.C.; Bombardelli, F.A.; Matos, J. Air Entrainment Onset in Skimming Flows on Steep Stepped Spillways: An Analysis. *J. Hydraul. Res.* **2014**, *52*, 375–385. [CrossRef]
15. Valero, D.; Bung, D.B.; Crookston, B.M. Energy Dissipation of a Type III Basin under Design and Adverse Conditions for Stepped and Smooth Spillways. *J. Hydraul. Eng.* **2018**, *144*, 04018036. [CrossRef]
16. Wan, H.; Li, R.; Gualtieri, C.; Yang, H.; Feng, J. Numerical Simulation of Hydrodynamics and Reaeration over a Stepped Spillway by the SPH Method. *Water* **2017**, *9*, 565. [CrossRef]
17. JCR, W.; Wray, A.; Moin, P. Eddies, Stream, and Convergence Zones in Turbulent Flows. *Stud. Turbul. Using Numer. Simul. Databases-11* **1988**, 193–207. Available online: https://ntrs.nasa.gov/api/citations/19890015184/downloads/19890015184.pdf (accessed on 28 November 2021).
18. Liu, C.; Gao, Y.; Tian, S.; Dong, X. Rortex—A New Vortex Vector Definition and Vorticity Tensor and Vector Decompositions. *Phys. Fluids* **2018**, *30*, 035103. [CrossRef]
19. Liu, C.; Wang, Y.; Yang, Y.; Duan, Z. New Omega Vortex Identification Method. *Sci. China Phys. Mech. Astron.* **2016**, *59*, 684711. [CrossRef]
20. Jeong, J.; Hussain, F. On the Identification of a Vortex. *J. Fluid Mech.* **1995**, *285*, 69–94. [CrossRef]
21. Wang, C.; Liu, Y.; Chen, J.; Zhang, F.; Huang, B.; Wang, G. Cavitation Vortex Dynamics of Unsteady Sheet/Cloud Cavitating Flows with Shock Wave Using Different Vortex Identification Methods. *J. Hydrodyn* **2019**, *31*, 475–494. [CrossRef]
22. Li, S.; Yang, J.; Liu, W. Estimation of Aerator Air Demand by an Embedded Multi-Gene Genetic Programming. *J. Hydroinformatics* **2021**, *23*, 1000–1013. [CrossRef]
23. Nóbrega, J.D.; Matos, J.; Schulz, H.E.; Canelas, R.B. Smooth and Stepped Spillway Modeling Using the SPH Method. *J. Hydraul. Eng.* **2020**, *146*, 04020054. [CrossRef]
24. Li, S.; Xie, Q.; Yang, J. Daily Suspended Sediment Forecast by an Integrated Dynamic Neural Network. *J. Hydrol.* **2022**, *604*, 127258. [CrossRef]
25. Felder, S.; Guenther, P.; Chanson, H. *Air-Water Flow Properties and Energy Dissipation on Stepped Spillways: A Physical Study of Several Pooled Stepped Configurations*; School of Civil Engineering: Brisbane, Australia, 2012.

26. Blocken, B.; Gualtieri, C. Ten Iterative Steps for Model Development and Evaluation Applied to Computational Fluid Dynamics for Environmental Fluid Mechanics. *Environ. Model. Softw.* **2012**, *33*, 1–22. [CrossRef]
27. Roache, P.J. Perspective: Validation—What Does It Mean? *J. Fluids Eng.* **2009**, *131*, 034503-3. [CrossRef]
28. Hirt, C.W.; Nichols, B.D. Volume of Fluid (VOF) Method for the Dynamics of Free Boundaries. *J. Comput. Phys.* **1981**, *39*, 201–225. [CrossRef]
29. Morovati, K.; Eghbalzadeh, A. Study of Inception Point, Void Fraction and Pressure over Pooled Stepped Spillways Using Flow-3D. *Int. J. Numer. Methods Heat Fluid Flow* **2018**, *28*, 982–998. [CrossRef]
30. Bayon, A.; Toro, J.P.; Bombardelli, F.A.; Matos, J.; López-Jiménez, P.A. Influence of VOF Technique, Turbulence Model and Discretization Scheme on the Numerical Simulation of the Non-Aerated, Skimming Flow in Stepped Spillways. *J. Hydro-Environ. Res.* **2018**, *19*, 137–149. [CrossRef]
31. Bombardelli, F.A.; Meireles, I.; Matos, J. Laboratory Measurements and Multi-Block Numerical Simulations of the Mean Flow and Turbulence in the Non-Aerated Skimming Flow Region of Steep Stepped Spillways. *Environ. Fluid Mech* **2011**, *11*, 263–288. [CrossRef]
32. Yakhot, V.; Orszag, S.A. Renormalization Group Analysis of Turbulence. I. Basic Theory. *J. Sci. Comput.* **1986**, *1*, 3–51. [CrossRef]
33. Celik, I.B.; Ghia, U.; Roache, P.J.; Freitas, C.J. Procedure for Estimation and Reporting of Uncertainty Due to Discretization in CFD Applications. *J. Fluids Eng.-Trans. ASME* **2008**, *130*, 078001.
34. Eça, L.; Hoekstra, M.; Roache, P. Verification of Calculations: An Overview of the 2nd Lisbon Workshop. In Proceedings of the 18th AIAA Computational Fluid Dynamics Conference, Miami, FL, USA, 25–28 June 2007; American Institute of Aeronautics and Astronautics: Reston, VA, USA.

Article

No Existence and Smoothness of Solution of the Navier-Stokes Equation

Hua-Shu Dou

Faculty of Mechanical Engineering and Automation, Zhejiang Sci-Tech University, Hangzhou 310018, China; huashudou@zstu.edu.cn

Abstract: The Navier-Stokes equation can be written in a form of Poisson equation. For laminar flow in a channel (plane Poiseuille flow), the Navier-Stokes equation has a non-zero source term ($\nabla^2 u(x, y, z) = F_x (x, y, z, t)$) and a non-zero solution within the domain. For transitional flow, the velocity profile is distorted, and an inflection point or kink appears on the velocity profile, at a sufficiently high Reynolds number and large disturbance. In the vicinity of the inflection point or kink on the distorted velocity profile, we can always find a point where $\nabla^2 u(x, y, z) = 0$. At this point, the Poisson equation is singular, due to the zero source term, and has no solution at this point due to singularity. It is concluded that there exists no smooth orphysically reasonable solutions of the Navier-Stokes equation for transitional flow and turbulence in the global domain due to singularity.

Keywords: Navier-Stokes equation; singularity; transitional flow; turbulence; Poisson equation

Citation: Dou, H.-S. No Existence and Smoothness of Solution of the Navier-Stokes Equation. *Entropy* **2022**, *24*, 339. https://doi.org/10.3390/e24030339

Academic Editors: Amine Ammar, Francisco Chinesta, Rudy Valetteand and Yinnian He

Received: 20 January 2022
Accepted: 24 February 2022
Published: 26 February 2022

Publisher's Note: MDPI stays neutral with regard to jurisdictional claims in published maps and institutional affiliations.

Copyright: © 2022 by the author. Licensee MDPI, Basel, Switzerland. This article is an open access article distributed under the terms and conditions of the Creative Commons Attribution (CC BY) license (https:// creativecommons.org/licenses/by/ 4.0/).

1. Introduction

In the past 50 years, researchers have conducted theoretical, experimental and direct numerical simulation (DNS) works on the Navier-Stokes equation and have shown that the flow field governed by this equation coincides well with the data on both the laminar flow and the turbulent flow. Therefore, people believe that the Navier-Stokes equation describes both the laminar flow and turbulence qualitatively and quantitatively. However, whether the three-dimensional (3D) incompressible Navier-Stokes equation has unique smooth (continuously differentiable) solutions is still not known [1,2].

Leray showed that the Navier-Stokes equations in three space dimensions always have a weak solution for velocity and pressure, with suitable growth properties [3], but the uniqueness of weak solutions of the Navier-Stokes equation is not demonstrated. Further, the existence of a strong solution (continuously differentiable) of the Navier-Stokes equations is still a challenge in the community of mathematics and physics, although much effort has been made around the world.

Dou and co-authors studied the origin of turbulence using the energy gradient theory [4–9] and discovered that there is velocity discontinuity in transitional flow and turbulence [9], which is a singularity of the Navier-Stokes equation. The singularity found theoretically is in agreement with the burst phenomenon in experiments. It was concluded that there exist no smooth and physically reasonable solutions of the Navier-Stokes equation at a high Reynolds number (beyond laminar flow) [9].

As is well known, the flow of viscous incompressible fluid is governed by the Navier-Stokes equation, which is a Poisson equation. The steady laminar flow is dominated by the Poisson equation with the source term of no vanishing. As observed in experiments and simulations, when the incoming laminar flow is disturbed by nonlinear disturbance, the velocity profile is distorted at a sufficient high Reynolds number. In the distorted flow, there may be some points on the velocity profile where the source term becomes zero, which form singularities of the corresponding Poisson equation. The existence of these singular points may lead to no solution of the Navier Stokes equation.

Singularity of the Navier-Stokes equation has received extensive study, owing to its importance in partial differential equations and turbulence [1]. In the literature, there are two different types of singularities described. These singularities are both located off the solid walls. The first type is the one formed by the unbounded kinetic energy of fluid in the flow field [1,3]. The second type is defined at the location where the streamwise velocity of fluid is theoretically zero [9]. The formation mechanisms of these two kinds of singularities are completely different. The former is caused by local infinite acceleration of fluid, and finally blowing up takes place. The latter is resulted from the variation of the velocity profile caused by disturbance in the flow field, which is the singularity of the Navier-Stokes equation itself at some location. This kind of singularity can only occur in viscous flow and does not occur in inviscid flow. In contrast, the first type of singularity may occur in inviscid flows [10]. It has been shown that the first type of singularity may be formed via reconnection of vortex rings in viscous flows [11,12].

In this study, the behavior of the Navier-Stokes equation in the Poisson equation form in transitional flow and turbulence is studied by analyzing the evolution of the velocity profile under finite disturbance, and the singular point of the Poisson equation is explored in the flow domain. No existence and smoothness of solution of the Navier-Stokes equation is concluded for transitional flow and turbulence.

Moffatt has restated the well-known Clay millennium prize problem essentially as this [13]: "can any initially smooth velocity field offinite energy in an incompressible fluid become singularat finite time under Navier-Stokes evolution?" The answer from the reasoning in present study is certainly, if the Reynolds number is sufficiently high and the disturbance is sufficiently large to lead to velocity deficit.

2. Stability and Turbulent Transition of Plane Poiseuille Flow

The three-dimensional laminar flow between two parallel walls is as shown in Figure 1 (plane Poiseuille flow). The width of space between two plates in the spanwise direction is infinite. The height in wall-normal direction between the two plates is 2h. The wall is set as the no-slip condition. The incoming flow is a laminar velocity profile. The downstream boundary is set as the Neumann boundary condition. The exact solution of the velocity for the laminar flow is a parabolic velocity distribution along the height for Newtonian fluid [14]. This smooth velocity distribution is placed in the flow field as the initial condition. Then, we observe the variation of the velocity distribution with time under finite disturbances, as in simulations and experiments [15–18].

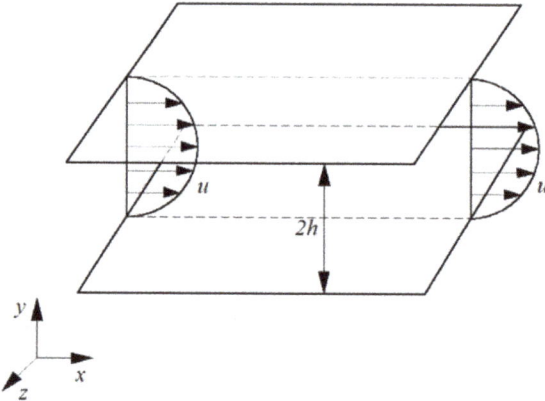

Figure 1. Plane Poiseuille flow between two parallel plates with boundary conditions and initial conditions.

With the flow development from the interaction of the base flow with the disturbance, the velocity profile can be modified, depending on the Reynolds number and the disturbance, as in simulations and experiments [15–18]. For the incoming laminar velocity profile (Figure 2a) at a sufficiently high Reynolds number, after the velocity profile is distorted, inflection point appears first (point A in Figure 2b), and then a section with positive second derivatives appears on the downstream velocity profile (section A–B in Figure 2c). The velocity profile with positive second derivatives may play an important role in the formation of singularity. Three features of the streamwise velocity profile are shown in Figure 2.

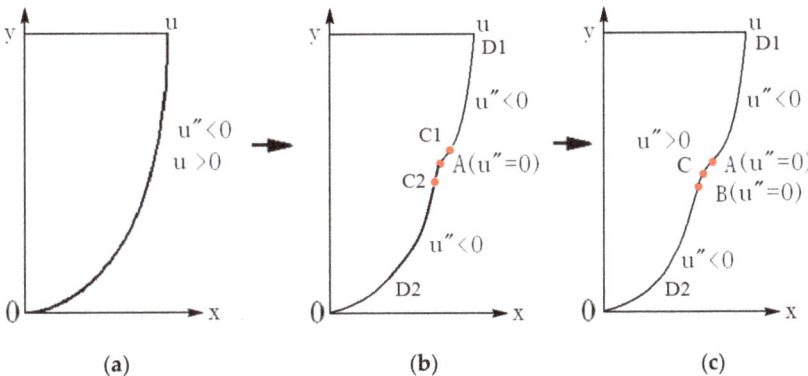

Figure 2. (a) Velocity profile of laminar flow; (b) an inflection point appears on the velocity profile indicated by A, and the second derivative of velocity $u'' < 0$, except point A; (c) the second inflection point B is produced after the first inflection point A, and a section of $u'' > 0$ appears on the velocity profile (A–B section). Here, u'' stands for the second derivative of the velocity to the direction normal to the wall, $\partial^2 u/\partial y^2$. In the figure, A: inflection point; B: second inflection point; D1: upper part; D2: lower part; C1: point between A and D1 in (b); C2: point between A and D2 in (b); C: point between A and B in (c).

Numerical simulations and experimental data show that when the laminar flow is disturbed, the velocity profile will change, and some positions of the velocity profile will be distorted. The results of theoretical analysis on plane Poiseuille flow by Dou show that the basic flow has the maximum ability to amplify the disturbance at $y/h = \pm 0.58$, and the velocity distortion is the largest there [4,5]. Numerical calculations and experiments have shown that the place where the maximum disturbance appears and the velocity profile change first occurs is at $y/h = \pm 0.58$ [15], where the velocity profile shows an inflection point. These results confirmed the analytical results by Dou and co-authors [4–9]. However, there is little change in the velocity profile at the center line and near the two walls in the early stage of disturbance amplification in plane Poiseuille flow.

Dou proved with the energy gradient theory that when there is an inflection point on the velocity profile, discontinuity (negative spike) of streamwise velocity occurs in the temporal evolution under disturbance [9], which is in agreement with simulations and experiments. A model for the velocity distribution at the discontinuity was proposed as shown in Figure 3, which occurs immediately after the inflection point is formed on the velocity profile.

Leray did pioneering work on the weak solution of the Navier-Stokes equations [3]. Foias et al. summarized that [1]: "Leray speculated that turbulence is due to the formation of point or 'line vortices' on which some component of the velocity becomes infinite." "Even today, despite much effort, Leray's conjecture concerning the appearance of singularities in 3-dimensional turbulent flows has been neither proved nor disproved".

As far as we know, the singularity in turbulence conjectured by Leray is never found in experiments and simulations (the first type of singularity mentioned in the Introduction). In contrast, the second type of singularity (zero streamwise velocity off the solid wall) is confirmed by experiments and simulation results [19–22], as described in [9]. The aim of present study is an alternative approach to achieve the same conclusion as that in our previous work [9]: that there exist no smooth and physically reasonable solutions of the Navier-Stokes equation for transitional flow and turbulence in the global domain due to singularity (for pressure driven flows).

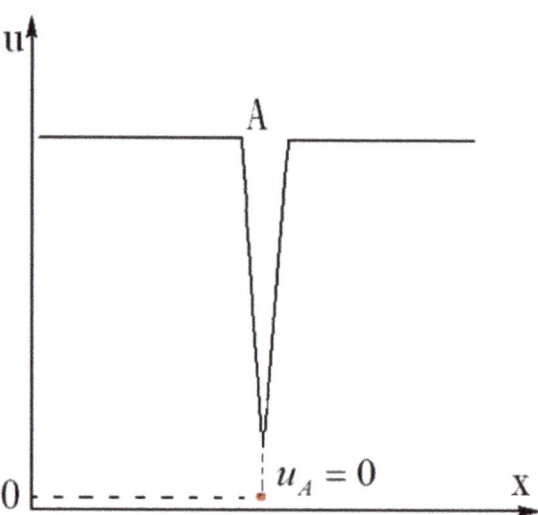

Figure 3. Streamwise velocity distribution under finite disturbance for high Reynolds number flow (transitional flow), showing the singular point (velocity discontinuity) in the vicinity of the inflection point A [9].

3. Navier-Stokes Equation in Form of Poisson Equation

3.1. Navier-Stokes Equation: Poisson Equation

The continuity and the unsteady momentum equation (Navier-Stokes equation) for incompressible fluid can be written as follows [14]:

$$\nabla \cdot \mathbf{u} = 0 \tag{1}$$

$$\rho\left(\frac{\partial \mathbf{u}}{\partial t} + \mathbf{u} \cdot \nabla \mathbf{u}\right) = -\nabla p + \mu \nabla^2 \mathbf{u} + \mathbf{f} \tag{2}$$

where $\mathbf{u}$ is the velocity vector, p is the static pressure, ρ is the fluid density, μ is the dynamic viscosity, and $\mathbf{f}$ is the gravitational force.

For the pressure-driven flow between two parallel plane walls (Figure 1), the wall boundary condition is no-slip:

$$\mathbf{u} = 0 \tag{3}$$

Rewriting Equation (2), we have

$$\nabla^2 \mathbf{u} = \frac{1}{\nu}\left(\frac{\partial \mathbf{u}}{\partial t} + \mathbf{u} \cdot \nabla \mathbf{u}\right) + \frac{1}{\mu}\nabla p - \frac{1}{\mu}\mathbf{f} \tag{4}$$

where $\nu = \mu/\rho$ is the kinematic viscosity.

Equation (4) is a form of Poisson equation and can be written as follows in Cartesian coordinates:

$$\frac{\partial^2 u}{\partial x^2} + \frac{\partial^2 u}{\partial y^2} + \frac{\partial^2 u}{\partial z^2} = F_x(x,y,z,t) \tag{5}$$

$$\frac{\partial^2 v}{\partial x^2} + \frac{\partial^2 v}{\partial y^2} + \frac{\partial^2 v}{\partial z^2} = F_y(x,y,z,t)$$

$$\frac{\partial^2 w}{\partial x^2} + \frac{\partial^2 w}{\partial y^2} + \frac{\partial^2 w}{\partial z^2} = F_z(x,y,z,t)$$

where

$$F_x(x,y,z,t) = \frac{1}{\nu}\left(\frac{\partial u}{\partial t} + u\frac{\partial u}{\partial x} + v\frac{\partial u}{\partial y} + w\frac{\partial u}{\partial z}\right) + \frac{1}{\mu}\frac{\partial p}{\partial x} - \frac{1}{\mu}f_x \tag{6}$$

$$F_y(x,y,z,t) = \frac{1}{\nu}\left(\frac{\partial v}{\partial t} + u\frac{\partial v}{\partial x} + v\frac{\partial v}{\partial y} + w\frac{\partial v}{\partial z}\right) + \frac{1}{\mu}\frac{\partial p}{\partial y} - \frac{1}{\mu}f_y$$

$$F_z(x,y,z,t) = \frac{1}{\nu}\left(\frac{\partial w}{\partial t} + u\frac{\partial w}{\partial x} + v\frac{\partial w}{\partial y} + w\frac{\partial w}{\partial z}\right) + \frac{1}{\mu}\frac{\partial p}{\partial z} - \frac{1}{\mu}f_z$$

where u, v, and w are the velocity components in the x, y, and z direction, respectively.

3.2. Reduced Form of Navier-Stokes Equation: Laplace Equation

When the source term in Equation (5) becomes zero in the domain ($-h \leq y \leq +h$),

$$F_x(x,y,z,t) = 0,\ F_y(x,y,z,t) = 0,\ F_z(x,y,z,t) = 0 \tag{7}$$

The Poisson Equation (5) reduces to the Laplace equation form,

$$\frac{\partial^2 u}{\partial x^2} + \frac{\partial^2 u}{\partial y^2} + \frac{\partial^2 u}{\partial z^2} = 0,\ \frac{\partial^2 v}{\partial x^2} + \frac{\partial^2 v}{\partial y^2} + \frac{\partial^2 v}{\partial z^2} = 0,\ \frac{\partial^2 w}{\partial x^2} + \frac{\partial^2 w}{\partial y^2} + \frac{\partial^2 w}{\partial z^2} = 0 \tag{8}$$

The solution of the Laplace equation in Equation (8) for plane Poiseuille flow with the no-slip boundary condition at the upper and bottom walls is

$$u(x,y,z) = 0,\ v(x,y,z) = 0,\ w(x,y,z) = 0 \tag{9}$$

This means that the fluid is static.

3.3. Solution of Navier-Stokes Equation for Steady Laminar Flow

For the steady parallel laminar flow in plane Poiseuille flow shown as in Figure 1, the pressure gradient in the x direction is

$$\partial p / \partial x \neq 0 \tag{10}$$

Then, the source term in Equation (5) is

$$F_x(x,y,z) \neq 0\ (-h \leq y \leq +h) \tag{11}$$

The solution for the Poisson Equation (5) with the no-slip boundary condition is

$$u(x,y,z) \neq 0\ (-h < y < +h) \tag{12}$$

$$u(x,y,z) = 0\ (y = \pm h)$$

For pressure-driven flow governed by Equation (5), the non-zero solution of velocity is that the source term must not be zero in the Poisson equation (Navier-Stokes equation). However, this conclusion is not true for plane Couette flow (shear driven flow), and a non-zero solution of velocity does not require a non-zero source term in the Poisson equation (Navier-Stokes equation).

The solution of Equation (5) with a source term $F_x(x,y,z,t) = 0$ at any location in the domain ($-h < y < +h$) makes no sense for the studied plane Poiseuille problem, and the position with $F_x(x,y,z,t) = 0$ would be a singular point of solution for Equation (12).

As is well known, for steady laminar flow at a low Reynolds number, the solution of Equation (4) or (5) for the parallel flow between two parallel plates (Figure 1) is as follow if the origin of the coordinates is fixed at the centerline [14]

$$u(y) = -\frac{\partial p}{\partial x}\frac{h^2}{2\mu}\left(1 - \frac{y^2}{h^2}\right) \tag{13}$$

$$v = 0$$

$$w = 0$$

For this solution at low Reynolds number, there is no singular point in such a laminar flow, and the flow is smooth and stable.

4. Solution of Navier-Stokes Equation for Transitional and Turbulent Flows

For the laminar flow at a higher Reynolds number in the plane Poiseuille flow configuration, once a disturbance is introduced, the flow becomes time-dependent and three-dimensional. As such, the streamwise velocity is distorted downstream due to the effect of nonlinear disturbance interaction (Figure 2b,c). The velocity components in this three-dimensional flow are $v(x,y,z) \neq 0$ and $w(x,y,z) \neq 0$, but $u >> v$ and $u >> w$. In the transitional flow and turbulent flow, at the position of turbulence "burst", the velocity components v and w may become large, which are the same order of magnitude as the streamwise component u, but the magnitudes of v and w are still much smaller than the streamwise velocity u, as found from previous experiments [19].

4.1. Strategy to Solve the Problem

As mentioned before, the existence and smoothness of the Navier-Stokes solution are still unknown. If we find the singularity in the flow field that makes the solution nonexistent in transitional flow and turbulence, we can provide counter evidence to the existence and smoothness of the Navier-Stokes equation.

As discussed for Equation (5), for the steady laminar flow between two parallel plates, $\nabla^2 u(x,y,z) = F_x(x,y,z)$ with $F_x(x,y,z) \neq 0$, and the solution within the domain is $u(x,y,z) \neq 0$, except at the wall boundaries. There is no singularity in the basic flow here.

In transitional flow, under the nonlinear interaction of disturbance, the velocity profile is distorted at a high Reynolds number, and the distorted velocity profile may produce singularity in temporal evolution. For example, at some point (like the inflection point on velocity profile, which will be shown later) on the velocity profile, the Laplace operator may instantaneously become zero, $\nabla^2 u(x,y,z) = 0$, so that $F_x(x,y,z) = 0$. At this point, the solution to satisfy the governing equation $\nabla^2 u(x,y,z) = 0$ and the boundary condition $u = 0$ at the wall is $u = 0$.

As shown in Figure 2b,c, the velocity of the given distorted velocity profile is $u \neq 0$ except at the wall, while the solution at the inflection point from the governing equation and the boundary condition is $u = 0$ in a time-dependent flow. Thus, this point is a singular point of the Navier-Stokes equation (Equation (5)). Since there exists a singular point in the flow field, there is no smooth solution of the Navier-Stokes equation.

When there is such a singular point, the Navier-Stokes equation has no solution at the singular point. In the transitional flow, we can find such singularity by analyzing the evolution of the velocity profile.

Under the condition at sufficiently high Re and finite disturbance, two types of velocity profiles can be produced, at least as shown in Figure 2b,c. After carrying out analysis, we can always find the point where $\nabla^2 u(x,y,z) = 0$ on these two types of velocity profiles. At such a point, $F_x(x,y,z,t) = 0$, which becomes the singularity of the Poisson equation (Equation (5)).

4.2. Finding the Singular Point on the Velocity Profile

The velocity profile of the u component under a disturbance expressed by Equation (5) is shown in Figure 2b,c, while the flow is actually three-dimensional at the transitional Reynolds number. At the inflection point (A in Figure 2b and A and B in Figure 2c), $\frac{\partial^2 u}{\partial y^2} = 0$, but $\frac{\partial^2 u}{\partial x^2} + \frac{\partial^2 u}{\partial z^2}$ may not be zero. However, in the vicinity of the inflection point on the velocity profile, we can always find the location where $\frac{\partial^2 u}{\partial y^2}$ and $\frac{\partial^2 u}{\partial x^2} + \frac{\partial^2 u}{\partial z^2}$ have identical magnitude and have opposite signs, since the gradient of flow variables in the y direction is much larger than those in the x and z directions. Thus, we can have $\nabla^2 u(x,y,z) = 0$ at such a location.

In the following, the singular point that makes the equation $\nabla^2 u(x,y,z) = 0$ established in 3D flows is explored.

(a) When the incoming flow is a laminar velocity profile (two-dimensional), $u = u(y)$, $v = 0$, $w = 0$, we have $\frac{\partial^2 u}{\partial y^2} < 0$, $\frac{\partial^2 u}{\partial x^2} = 0$, $\frac{\partial^2 u}{\partial z^2} = 0$, and $\frac{\partial^2 u}{\partial x^2} + \frac{\partial^2 u}{\partial y^2} + \frac{\partial^2 u}{\partial z^2} < 0$ (Figure 2a).

(b) When the velocity is disturbed by finite disturbance, the middle part of the velocity profile is disturbed greatly (becoming three-dimensional), and this part first deforms. Here, it reaches $\frac{\partial^2 u}{\partial y^2} = 0$ first in the middle part, forming the inflection point A, while the upper and lower parts of the velocity profile do not change much (Figure 2b). In the following, we will discuss these as two different cases, respectively.

(c) If $\frac{\partial^2 u}{\partial x^2} + \frac{\partial^2 u}{\partial z^2} > 0$ at point A (noting $\frac{\partial^2 u}{\partial y^2} = 0$ at point A), then $\frac{\partial^2 u}{\partial x^2} + \frac{\partial^2 u}{\partial y^2} + \frac{\partial^2 u}{\partial z^2} > 0$.

Observing Figure 2b, in the top and bottom parts of the velocity profile, D1 and D2 locations, the disturbance is small, and there still exist $\frac{\partial^2 u}{\partial x^2} \approx 0$, $\frac{\partial^2 u}{\partial z^2} \approx 0$, $\frac{\partial^2 u}{\partial y^2} < 0$, $\left|\frac{\partial^2 u}{\partial y^2}\right| \gg \left|\frac{\partial^2 u}{\partial x^2} + \frac{\partial^2 u}{\partial z^2}\right|$; thus, $\frac{\partial^2 u}{\partial x^2} + \frac{\partial^2 u}{\partial y^2} + \frac{\partial^2 u}{\partial z^2} < 0$. Then, it is necessary that there exist C1 and C2 points between A and D1 and between A and D2, respectively, which makes $\nabla^2 u(x,y,z) = 0$.

(d) If $\frac{\partial^2 u}{\partial x^2} + \frac{\partial^2 u}{\partial z^2} < 0$ at point A (noting $\frac{\partial^2 u}{\partial y^2} = 0$ at point A), then $\frac{\partial^2 u}{\partial x^2} + \frac{\partial^2 u}{\partial y^2} + \frac{\partial^2 u}{\partial z^2} < 0$.

Observing Figure 2c, the velocity profile develops, reaching the stage of section A–B where $\frac{\partial^2 u}{\partial y^2} > 0$ (section A–B), while $\frac{\partial^2 u}{\partial x^2} + \frac{\partial^2 u}{\partial z^2} < 0$. With the evolution of the velocity profile, the value of $\frac{\partial^2 u}{\partial y^2}$ at C point increases gradually from zero to positive. When the magnitude of $\frac{\partial^2 u}{\partial y^2}$ equals that of $\frac{\partial^2 u}{\partial x^2} + \frac{\partial^2 u}{\partial z^2}$, then we have $\nabla^2 u(x,y,z) = 0$ at one point in the A–B section.

Therefore, when an inflection point is formed on the velocity profile, a section with a positive value of $\frac{\partial^2 u}{\partial y^2}$ will be formed inevitably, with further evolution under a sufficiently large disturbance at a sufficiently high Reynolds number. As long as the positive value of $\frac{\partial^2 u}{\partial y^2}$ is sufficiently large at this section (A–B) on the velocity profile and is able to offset the value of $\frac{\partial^2 u}{\partial x^2} + \frac{\partial^2 u}{\partial z^2}$, there always exists a location within the A–B section with $\nabla^2 u(x,y,z) = 0$.

4.3. Solution with Variation of Source Term

The following analysis can be made for various magnitudes of the source term in Equation (5):

(1) $F_x(x,y,z,t) = 0$, Equation (5) becomes the Laplace equation, Equation (7), and the solution is $u(x,y,z) = 0$, $v(x,y,z) = 0$, and $w(x,y,z) = 0$.

(2) $F_x(x,y,z,t)$ is small, i.e., Re is low, the nonlinear disturbance effect is small in the base laminar flow, and the velocity profile is not distorted downstream. Finally, the disturbance is damped, and the flow stays laminar.

(3) $F_x(x,y,z,t)$ is large, i.e., Re is high, the nonlinear disturbance effect is larger, and the velocity profile is distorted downstream, which leads to a transitional flow. The inflection point or kink appears on the velocity profile in the transitional flow. There is a position in the vicinity of the inflection point (A) where $\nabla^2 u(x,y,z) = 0$ is established (Figure 2b,c). This position becomes the singular point of the Poisson equation (Equation (5)) due to $F_x(x,y,z,t) = 0$ at this point. Figure 4 shows the schematic of the solution strategy of the disturbed flow.

In Figure 4a, the base flow is defined by the Poisson equation (Navier–Stokes), $\nabla^2 u(x,y,z) = F_x(x,y,z,t)$, and the source term is not zero, $F_x(x,y,z,t) \neq 0$. The solution of the governing equation with the wall no-slip boundary conditions is $u(x,y,z) \neq 0$, except at the walls.

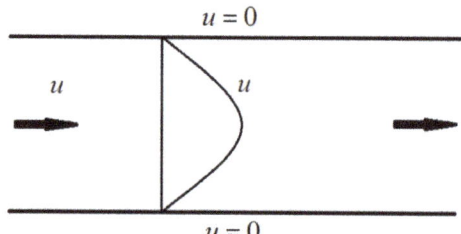

(a) Base flow is defined.

$\left. \begin{array}{l} \nabla^2 u(x,y,z) = f(x,y,z,t) \\ f(x,y,z,t) \neq 0 \end{array} \right\} \begin{array}{l} \text{Solution} \\ u(x,y,z) \neq 0 \\ u(x,y,z) = 0 \text{ at walls.} \end{array}$

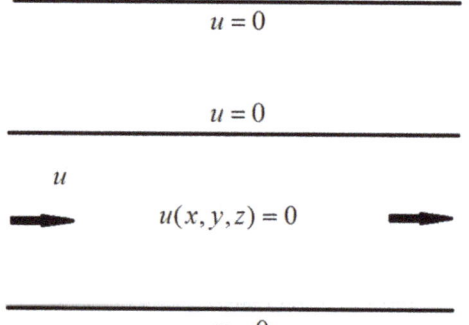

(b) Disturbed Flow.

$\nabla^2 u(x,y,z) = f(x,y,z,t)$, $f(x,y,z,t) \neq 0$,

but at point A,

$\nabla^2 u(x,y,z) = 0$, $f(x,y,z,t) = 0$,

which is singular.

There is no solution at point A.

(c) Solution of flow field with $\nabla^2 u(x,y,z) = 0$.

For flow field controlled by $\nabla^2 u(x,y,z) = 0$, and wall no-slip boundary condition, the solution is $u(x,y,z)=0$.

Figure 4. Solution of streamwise velocity showing that there is a singular point for the disturbed velocity distribution.

In Figure 4b, after the base flow is disturbed, the velocity profile is distorted locally. The governing equation is still the Poisson equation (Navier–Stokes), $\nabla^2 u(x,y,z) = F_x(x,y,z,t)$, and the source term is still not zero, $F_x(x,y,z,t) \neq 0$. However, there is an exception at the inflection point A or its neighborhood, $\nabla^2 u(x,y,z) = 0$ (i.e., $F_x(x,y,z,t) = 0$), which is singular in the flow field. Thus, there is no solution for Equation (5) at point A.

In Figure 4c, for flow field controlled by $\nabla^2 u(x,y,z) = 0$ and the wall no-slip boundary condition, the solution of flow field is $u(x,y,z) = 0$. Thus, the value of the streamwise velocity at point A in Figure 4b should be zero, as discussed for Equations (7)–(9).

Thus, after inflection points are formed in Figure 4b, the velocity at point A in Figure 4b will theoretically become zero immediately at the next moment in the temporal evolution. This is shown in Figure 5. Because of the viscosity of fluid, the velocity at point A will not be absolutely zero, but spikes are produced, as shown in Figure 3. Simultaneously, fluctuations of the velocity components v and w as well as the pressure p are produced, which follow the conservation of the total mechanical energy before and after the spike generation. At the inflection point or its neighborhood, where the spike is produced, the fluid element is compressed in the streamwise direction, and thus it is stretched in wall-normal and spanwise directions. Since the mainstream velocity decreases, the pressure will increase at the said singular location. Therefore, the fluctuation of u is firstly negative, and those of v, w and p are positive, at the singular point. These variations of fluctuations of velocity components and pressure are in agreement with the experimental results of turbulent burst in plane Poiseuille flow [19,23]. The feature of positive pressure maximum associated with the burst of turbulence has also been found in the boundary layer flow on flat plates [24–26].

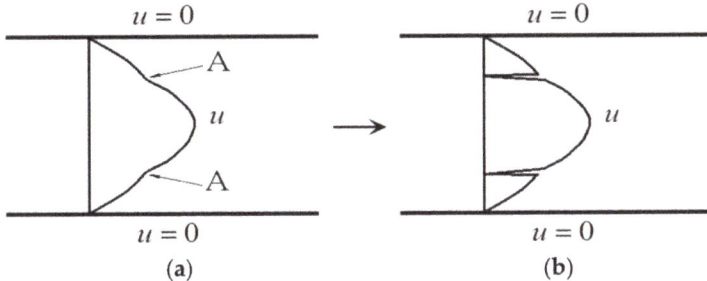

Figure 5. Temporal evolution of velocity profile in transitional flow. (**a**) Velocity profile with inflection points. (**b**) Singular points appear theoretically with $u = 0$ at the locations of zero source term.

Swearingen et al. found through experiments and simulations for wall-bounded flow that the turbulence production events are preceded by an inflectional velocity profile [27]. In bounded transitional flows, this unstable profile produces velocity fluctuations in the streamwise direction and in the other two directions. Figure 5 provides a theoretical interpretation for the generation of fluctuations in wall-bounded transitional flows (pressure driven flow). The result shown in Figure 5 is in agreement with the experimental and simulation results in [19–22].

In Figure 4b, since point A is singular, the velocity is not differentiable at this position (as shown in Figures 3 and 5). Therefore, there is no smooth solution of the Navier-Stokes equation (Equation (5)) at point A in the transitional flow (Figure 4b).

(4) For fully developed turbulent flow, $F_x(x,y,z,t)$ is large, i.e., Re is very high. The velocity profile is heavily distorted by the vortex overlap to the streamwise velocity profile [9], and velocity profiles with an inflectional point or kink are formed, which leads to $\nabla^2 u(x,y,z) = 0$ (i.e., $F_x(x,y,z,t) = 0$) at these points. As such, there is a lot of singular points of Equation (5), as expressed in Figure 2b,c. Therefore, there exist no continuous smooth solutions of Equation (5) for turbulent flow in the global domain.

Fletcher has discussed the characteristics of the general Poisson equation, where the source term can be set up as any value [28]. That is, there is no limitation on the value of the source term, and the zero source term can be defined anywhere in the flow field. For example, for the thermal conduction problem between two parallel plates, the source function of the resulting Poisson equation can be taken as any value (including zero value), and the solution of temperature function has no singularity within the domain.

For plane Poiseuille flow, the governing equation (Navier-Stokes equation) can be written as a form of Poisson equation, as in Equation (5). According to the Navier-Stokes equation and the boundary conditions of plane Poiseuille flow, the source term of the Poisson equation is not arbitrary and it must not be zero. Thus, the zero value of the source term of Equation (5), if any, is not defined in the domain. At any position in the flow field, as long as the source term is zero, it constitutes the singularity of the Navier-Stokes equation, which makes the equation have no solution.

5. Significance of the Singularity in Turbulence

In the transition of laminar flow to turbulence, the singularity at the inflection point (or its neighborhood) results in a "burst", as observed by previous experiments [9]. The burst is the origin of turbulence generation and production of turbulent stresses. In other words, the flow relieves this singularity by a "burst" at the inflection point, and the singularity is converted into turbulent fluctuations at the said position.

In present study, singularity on the velocity profile in finite time for the incompressible Navier-Stokes equations is both mathematical and physical. In mathematics, this singularity of the Navier-Stokes equation occurs at (or near) the inflection point, which makes the equation not differentiable at this position. In physics, this singularity leads to a "burst" as well as fluctuations of streamwise velocity and other velocity components, i.e., turbulence if the Reynolds number is sufficiently high. The mechanical energy of the mainstream flow is transmitted to turbulent fluctuations via this singularity. The singularity is also the reason why turbulence cannot be repeated exactly.

In physical fluid flows in the laboratory, the flow is three-dimensional under finite disturbance, rather than two-dimensional. Further, two-dimensional disturbance isn't able to induce this type of singularity, since the fluid element is subjected to both shear and extension at the singularity. In other words, there is also stretch of spanwise vorticity at the singular point.

As discussed before, at the singular point (inflection point), the streamwise velocity is theoretically zero. In order to conserve the total mechanical energy, the pressure at this point reaches its maximum. In the near upstream of the singular point, the disturbance is already three-dimensional (even though the amplitude is small). At the singular point, the disturbance is largest, with that singularity leading to "explosive burst" where the amplitudes of fluctuations of both velocity and pressure are largest. The pressure maximum at the singular point produces a pressure wave that spreads as its elliptical property.

Finally, it is pointed out that since the studied singularity of the Navier-Stokes equation is caused by a vanishing viscous term in the Navier-Stokes equation at the inflection point of the velocity profile in the flow domain (Laplace operator is zero), such singularity is never produced in inviscid flows governed by the Euler equation. This implies that turbulence, the properties of which are dissipative, with temporal bursting, with increasing resistance, and with self-sustained fluctuations, could not be generated in inviscid flows.

6. Conclusions

Solutions of the Navier-Stokes equation with the Poisson equation form are studied by analyzing the variation of the velocity profile versus the Re number and the disturbance. For the steady laminar flow between two parallel plates, the Poisson equation dominates the flow with the source term of no vanishing. For the laminar flow at a sufficiently high Reynolds number and under certain finite disturbance, the velocity profile is distorted downstream and an inflection point appears (or kink appearance). With the evolution of the velocity profile under finite disturbance, in the vicinity of the inflection point, it is found that there is always a position with $\nabla^2 u(x,y,z) = 0$ (i.e., $F_x(x,y,z,t) = 0$). This point is singular in the global domain for the Poisson equation (Navier-Stokes equation). At this kind of singular point, the flow variables are not differentiable. Therefore, there exist no smooth and physically reasonable solutions of the Navier-Stokes equation in transitional and turbulent flows.

It should be pointed out that the reasoning presented in this study is only for pressure-driven flows. For shear driven flow, the same conclusion on existence and smoothness of the solution of the Navier-Stokes equation can be obtained with the boundary conditions changed, but the work done by shear stress should be taken into account (this work will be published in separate paper). For shear-driven flows, the mechanisms of instability occurrence and turbulent transition have been studied, respectively, for plane Couette flow and Taylor-Couette flow in [7,8], where external work has been included.

The above conclusion confirmed the analysis results with the energy gradient theory in [9], which show occurrences of streamwise velocity suddenly stop and velocity discontinuity due to zero mechanical energy drop along the streamline. It was shown that the discontinuity of streamwise velocity forms the singularity of the Navier-Stokes equation.

Therefore, both approaches using energy gradient theory and Poisson equation analysis are consistent and show that there is a singular point in the vicinity of the inflection point on the velocity profile where the streamwise velocity is theoretically zero. Neither existence nor smoothness of the solution of the Navier-Stokes equation is demonstrated for transitional and turbulent flows.

Funding: This research received no external funding.

Institutional Review Board Statement: Not applicable.

Informed Consent Statement: Not applicable.

Data Availability Statement: The data are contained within the article.

Conflicts of Interest: The authors declare no conflict of interest.

References

1. Foias, C.; Manley, O.; Rosa, R.; Temam, R. *Navier-Stokes Equations and Turbulence*; Cambridge University Press: Cambridge, UK, 2004.
2. Fefferman, C.L. *Existence and Smoothness of the Navier-Stokes Equation*; Clay Mathematics Institute: Peterborough, NH, USA, 2000; pp. 1–6. Available online: http://www.claymath.org/sites/default/files/navierstokes.pdf (accessed on 23 February 2022).
3. Leray, J. Sur le mouvement d'un liquide visquex emplissent l'espace. *Acta Math. J.* **1934**, *63*, 193–248. [CrossRef]
4. Dou, H.-S. Energy gradient theory of hydrodynamic instability. In Proceedings of the Third International Conference on Nonlinear Science, Singapore, 30 June–2 July 2004. Available online: https://www.researchgate.net/publication/2147222 (accessed on 23 February 2022).
5. Dou, H.-S. Mechanism of flow instability and transition to turbulence. *Inter.J. Non-Linear Mech.* **2006**, *41*, 512–517. Available online: https://www.researchgate.net/publication/245215903 (accessed on 23 February 2022). [CrossRef]
6. Dou, H.-S. Physics of flow instability and turbulent transition in shear flows. *Inter. J. Phys. Sci.* **2011**, *6*, 1411–1425. Available online: https://www.researchgate.net/publication/2176288 (accessed on 23 February 2022).
7. Dou, H.-S.; Khoo, B.C. Investigation of turbulent transition in plane Couette flows using energy gradient method. *Advances in Appl.Math. and Mech.* **2011**, *3*, 165–180. Available online: https://www.researchgate.net/publication/2147221 (accessed on 23 February 2022). [CrossRef]
8. Dou, H.-S.; Khoo, B.C.; Yeo, K.S. Instability of Taylor-Couette Flow between Concentric Rotating Cylinders. *Inter. J. Thermal.Sci.* **2008**, *47*, 1422–1435. Available online: https://www.researchgate.net/publication/222709642 (accessed on 23 February 2022). [CrossRef]
9. Dou, H.-S. Singularity of Navier-Stokes equations leading to turbulence, Advances in Applied Mathematics and Mechanics. *arXiv Prepr.* **2021**, *13*, 527–553. Available online: https://www.researchgate.net/publication/325464812 (accessed on 23 February 2022).
10. Beale, J.T.; Kato, T.; Majda, A.J. Remarks on the breakdown of smooth solutions for the 3D Euler equations. *Commun. Math. Phys.* **1984**, *94*, 61–66. [CrossRef]
11. Yao, J.; Hussain, F. On singularity formation via viscous vortex reconnection. *J. Fluid Mech.* **2020**, *888*, R2. [CrossRef]
12. Moffatt, H.K.; Kimura, Y. Towards a finite-time singularity of the Navier-Stokes equations. Part 2. Vortex reconnection and singularity evasion. *J. Fluid Mech.* **2019**, *870*, R1. [CrossRef]

13. Moffatt, H.K. Singularities in fluid mechanics. *Phys. Rev. Fluids* **2019**, *4*, 110502. [CrossRef]
14. Schlichting, H. *Boundary Layer Theory*, 7th Ed. ed; Springer: Berlin, Germany, 1979.
15. Nishioka, M.; Iida, S.; Ichikawa, Y. An experimental investigation of the stability of plane Poiseuille flow. *J. Fluid Mech.* **1975**, *72*, 731–751. [CrossRef]
16. Biringen, S. Final stages of transition to turbulence in plane channel flow. *J. Fluid Mech.* **1984**, *148*, 413–442. [CrossRef]
17. Sandham, N.D.; Kleiser, L. The late stages of transition to turbulence in channel flow. *J. Fluid Mech.* **1992**, *245*, 319–348. [CrossRef]
18. Luo, J.; Wang, X.; Zhou, H. Inherent mechanism of breakdown in laminar-turbulent transition of plane channel flows, Science in China Ser. G Physics. *Mech. Astron.* **2005**, *48*, 228–236. [CrossRef]
19. Luchikt, T.S.; Tiederman, W.G. Timescale and structure of ejections and bursts in turbulent channel flows. *J. Fluid Mech.* **1987**, *174*, 524–552.
20. Alfredsson, P.H.; Johansson, A.V. On the detection of turbulence-generating events. *J. Fluid Mech.* **1984**, *139*, 325–345. [CrossRef]
21. Schlatter, P.; Stolz, S.; Kleiser, L. Large-eddy simulation of spatial transition in plane channel flow. *J. Turbul.* **2006**, *7*, N33. [CrossRef]
22. Nishioka, M.; Asai, M.; Iida, S. Wall phenomena in the final stage of transition to turbulence. In *Transition and Turbulence*; Meyer, R.E., Ed.; Academic Press: New York, NY, USA, 1981; pp. 113–126.
23. Kim, J.; Moin, P.; Moser, R.D. Turbulence statistics in fully developed channel flow at low Reynolds number. *J. Fluid Mech.* **1987**, *177*, 133–166. [CrossRef]
24. Johansson, A.V.; Her, J.-Y.; Haritonidis, J. On the generation of high-amplitude wall-pressure peaks in turbulent boundary layers and spots. *J. Fluid Mech.* **1987**, *175*, 119–142. [CrossRef]
25. Bernard, P.S.; Thomas, J.M.; Handler, R.A. Vortex dynamics and the production of Reynolds stress. *J. Fluid Mech.* **1993**, *253*, 385–419. [CrossRef]
26. Ghaemi, S.; Scarano, F. Turbulent structure of high-amplitude pressure peaks within the turbulent boundary layer. *J. Fluid Mech.* **2013**, *735*, 381–426. [CrossRef]
27. Swearingen, J.D.; Blackwelder, R.F.; Spalart, P.R. Inflectional instabilities in the wall region of bounded turbulent shear flows. In Proceedings of the Summer Program; Center for Turbulence Research, Stanford Univercity: Stanford, CA, USA, 1987; pp. 291–295.
28. Fletcher, C.A.J. *Computational Techniques for Fluid Dynamics*; Springer: Berlin, Germany, 1991; Volume 1.

Article

Numerical Study on the Coagulation and Breakage of Nanoparticles in the Two-Phase Flow around Cylinders

Ruifang Shi [1], Jianzhong Lin [1,2,*] and Hailin Yang [1]

[1] State Key Laboratory of Fluid Power and Mechatronic System, Zhejiang University, Hangzhou 310027, China; 11924031@zju.edu.cn (R.S.); yanghailin@zju.edu.cn (H.Y.)

[2] Laboratory of Impact and Safety Engineering, Ningbo University, Ministry of Education, Ningbo 315201, China

* Correspondence: mecjzlin@public.zju.edu.cn; Tel.: +86-571-87952221

Abstract: The Reynolds averaged N-S equation and dynamic equation for nanoparticles are numerically solved in the two-phase flow around cylinders, and the distributions of the concentration M_0 and geometric mean diameter d_g of particles are given. Some of the results are validated by comparing with previous results. The effects of particle coagulation and breakage and the initial particle concentration m_{00} and size d_0 on the particle distribution are analyzed. The results show that for the flow around a single cylinder, M_0 is reduced along the flow direction. Placing a cylinder in a uniform flow will promote particle breakage. For the flow around multiple cylinders, the values of M_0 behind the cylinders oscillate along the spanwise direction, and the wake region in the flow direction is shorter than that for the flow around a single cylinder. For the initial monodisperse particles, the values of d_g increase along the flow direction and the effect of particle coagulation is larger than that of particle breakage. The values of d_g fluctuate along the spanwise direction; the closer to the cylinders, the more frequent the fluctuations of d_g values. For the initial polydisperse particles with d_0 = 98 nm and geometric standard deviation σ = 1.65, the variations of d_g values along the flow and spanwise directions show the same trend as for the initial monodisperse particles, although the differences are that the values of d_g are almost the same for the cases with and without considering particle breakage, while the distribution of d_g along the spanwise direction is flatter in the case with initial polydisperse particles.

Keywords: nanoparticle two-phase flow; particle coagulation and breakage; flow around circular cylinders; particle distribution

1. Introduction

The particle-laden flow around cylinders has attracted the attention of many scholars because of its extensive industrial applications. The particle dispersion and distribution in the wake behind cylinders have been extensively studied in experimental and numerical simulations during the past decade. Zhou et al. [1] showed that the particle distribution was dependent on the Stokes number (St). The particles dispersed into the core regions of the vortex at small St values concentrated on the boundary of vortex at intermediate St values and assembled in the outer region of the vortex at large St values. Haddadi et al. [2] found that the hydrodynamic interaction between particles led to the exchange of particles between the wake area and the free stream. Jafari et al. [3] indicated that particle motion in the wake was strongly affected by vortex shedding, while the particle Brownian diffusion influenced the deposition rate of particles. Haddadi et al. [4] studied the suspension around the cylinder with a particle fraction of about 0.08 in the microchannel and found that particles could escape the wake area due to velocity fluctuations by increasing the particle number in the wake. There was also particle exchange between the wake area and the free stream. Huang et al. [5] indicated that the particle collection efficiency was positively related to the particle formation fraction, while the thermophoresis enhanced the impaction efficiency of particles by 1–2 orders. Jeong and Kim [6] indicated that the thermophorectic effect on the particles was obvious at small St values, while the deposition efficiency of particles was increased by increasing the temperature difference between the flow and the cylinder or by decreasing the ratio of thermal conductivity of the particle to the fluid. The particles with small St values follow the fluid in the upstream surface of the cylinder without collision, but move backward in the downstream direction. Gopan et al. [7] developed correlations for particle temperatures and impaction rates based on the flow and boundary conditions, as well as particle properties.

However, most of the above studies involved large particles, without considering the coagulation and breakage of particle clusters after coagulation. In practical applications, particle coagulation and breakage often occur, which affects the distribution of the particle concentration and sizein the wake behind cylinders. In addition, particle dispersion is dominant when the particle density is low, while it is necessary to consider the coagulation caused by particle collision when the particle density is high. Keita et al. [8] numerically studied the nanoparticle dispersion at Re = 9300. They found that the Brownian diffusion tended to concentrate the particles at the edges of the vortices, while the turbulence dispersed particles from the periphery to the core of the vortices. Tu and Zhang [9] experimentally studied the condensation of submicron- and nanoscale particles within Re = 5200–35,000. They found that the particle diameter downstream of the cylinder was larger than that upstream, and the total particle concentration and geometric mean diameter in the free stream werelarger and less than in the wake, respectively. Multi-cylinder alignment could be used to enhance particle coagulation. Liu et al. [10] numerically and experimentally studied the structural properties of the vortex generators affecting the particle coagulation. They showed that the optimal efficiency of the particle coagulation was about 16.42% for particles with sizes ranging from 15.7 to 850.0 nm at aflow velocity of 4.8 m/s. Particle collision and coagulation mainly occurred in the windward boundary layer of the vortex generator and at the longitudinal edges of the vortices. Kolsiet al. [11] numerically studied the effects of using double rotating cylinders and partly porous layers in the bifurcating channels on the hydrothermal performance and indicated that the proposed methods of heat transfer enhancement could be considered simultaneously for effective control of the thermal performance of those systems. Alsaberyet al. [12] investigated transient entropy generation and mixed convection due to a rotating hot inner cylinder within a square cavity with a flexible side wall and achieved the highest average heat transfer and global entropy generation rates for counter-clockwise rotation of the circular cylinder and lower values in terms of flexible wall deformation.

From the above, it can be seen that there are few studies on particle distribution while simultaneously considering the effects of particle coagulation and breakage, and research results for the two-phase flow of nanoparticles around multiple cylinders are also rare. In this work, a numerical simulation is carried out to study the distribution of the particle concentration and particle size in the two-phase flow of nanoparticles around a single cylinder and multiple cylinders. The flow of fluid around cylinders is selected because this kind of flow is the most common in practical applications. Meanwhile, the effects of particle coagulation and breakage, initial particle concentration, and size on the particle distribution are discussed.

2. Governing Equations

For the nanoparticle-laden flow around cylinders, as shown in Figure 1 the distance between two walls in the z direction is long enough and there is no velocity in the z direction, meaning that the change of the flow along the z direction can be ignored. Therefore, the three-dimensional flow can be reduced to flow the in x-y plane. The distribution of the particle concentration and size in the wake behind cylinders is closely related to the flow characteristics. In Figure 1a, the cylinder diameter is D = 50 mm and the geometric center of the cylinder is used as the coordinate origin. Coordinates x, y, and z represent the flow, vertical direction, and span direction, respectively. In Figure 1b, the cylinder diameter is $D = 10$ mm; the flow direction positions of the four columns of cylinders are $x = 0.05, 0.09, 0.13$, and 0.17 m, respectively; and the spanwise positions of the five rows of cylinders are $y = \pm 0.02, 0$, and ± 0.04 m, respectively. The Reynolds number of the flow is defined as $Re = UD/\nu$.

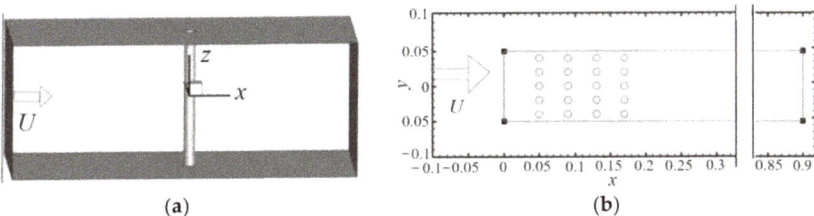

Figure 1. Nanoparticle-laden flows: (**a**) flow around a cylinder; (**b**) flow around 20 cylinders.

2.1. Fluid Flow

For the incompressible flow, the continuity and the Reynolds averaged N-S equations are:

$$\frac{\partial u_i}{\partial x_i} = 0 \tag{1}$$

$$\rho \left(\frac{\partial u_i}{\partial t} + u_k \frac{\partial u_i}{\partial x_k} \right) = -\frac{\partial p}{\partial x_i} + \frac{\partial}{\partial x_j}\left(\mu \frac{\partial u_i}{\partial x_j}\right) + \frac{\partial}{\partial x_j}\left(-\rho \overline{u_i' u_j'}\right) \tag{2}$$

where u_i is the mean velocity, ρ is the density, p is the pressure, μ is the fluid viscosity, and $\overline{u_i' u_j'}$ is the Reynolds stress, which is modeled by:

$$-\rho \overline{u_i' u_j'} = \rho \nu_t \left(\frac{\partial u_i}{\partial x_j} + \frac{\partial u_j}{\partial x_i} \right) - \frac{2}{3}\rho k \delta_{ij} \tag{3}$$

in which $\nu_t = C_\mu k^2/\varepsilon$, where C_μ is a function of the average strain rate, and k and ε are the turbulent kinetic energy and dissipation rate, respectively. The standard k-ε turbulence model is selected here because it is suitable for flow with Reynolds numbers in the range of $300 < Re < 3 \times 10^5$ and is not directly affected by the wall. The transport equations of k and ε are:

$$\frac{\partial k}{\partial t} + u_j \frac{\partial k}{\partial x_j} = \frac{\partial}{\partial x_j}\left[\left(\nu + \frac{\nu_t}{\sigma_k}\right)\frac{\partial k}{\partial x_j}\right] - \rho_f \overline{u_i' u_j'}\frac{\partial u_i}{\partial x_j} - \varepsilon \tag{4}$$

$$\frac{\partial \varepsilon}{\partial t} + u_j \frac{\partial \varepsilon}{\partial x_j} = \frac{\partial}{\partial x_j}\left[\left(\nu + \frac{\nu_t}{\sigma_\varepsilon}\right)\frac{\partial \varepsilon}{\partial x_j}\right] + \frac{\varepsilon}{k}\nu_t C_{\varepsilon 1}\left(-\rho_f \overline{u_i' u_j'}\frac{\partial u_i}{\partial x_j}\right) - \rho_f C_{\varepsilon 2}\frac{\varepsilon^2}{k} \tag{5}$$

where $C_\mu = 0.09$, $C_{\varepsilon 1} = 1.44$, $C_{\varepsilon 2} = 192$, $\sigma_k = 1.0$, and $\sigma_\varepsilon = 1.3$.

2.2. General Dynamic Equation for Nanoparticles

The instantaneous general dynamic equation for nanoparticles considering the convection, diffusion, and particle coagulation and breakage is [13,14]:

$$\frac{\partial n(v,t)}{\partial t} + u_i \frac{\partial n(v,t)}{\partial x_i} - \frac{\partial}{\partial x_i}\left[D_T \frac{\partial n(v,t)}{\partial x_i}\right] = \frac{1}{2}\int_0^v \beta(v_1, v - v_1)n(v_1,t)n(v-v_1,t)dv_1$$
$$-\int_0^\infty \beta(v_1,v)n(v,t)n(v_1,t)dv_1 + \int_v^\infty a(v_1)b(v|v_1)n(v_1)dv_1 - a(v)n(v), \tag{6}$$

where n is the spatial distribution of the number of particles with volume v at time t, and D_T is the turbulent diffusion coefficient, which is approximated by the turbulent viscosity of the fluid [15]. The first two terms on the right hand side of Equation (6) are the generation and disappearance of particles with volume v caused by coagulation, while $\beta(v_1,v)$ is the kernel function of coagulation of the two particles with volume v and volume v_1 and consists of two parts, i.e., $\beta = \beta_B + \beta_T$, where β_B is the coagulation kernel caused by Brownian motion [16]:

$$\beta_B = \frac{2KT}{3\mu}\left(\frac{1}{v^{1/3}} + \frac{1}{v_1^{1/3}}\right)\left(v^{1/3} + v_1^{1/3}\right) + \frac{2KT}{3\mu}\frac{1.591\lambda}{(3/4\pi)^{1/3}}\left(\frac{1}{v^{2/3}} + \frac{1}{v_1^{2/3}}\right)\left(v^{1/3} + v_1^{1/3}\right) \qquad (7)$$

where K is the Boltzmann constant, T is the temperature, and λ is the average free path of gas molecules. Here, β_T is the coagulation kernel caused by turbulent shear [17]:

$$\beta_T = \sqrt{\frac{3}{10\pi}}\left(v^{1/3} + v_1^{1/3}\right)^3\left(\frac{\varepsilon}{\nu}\right)^{1/2} \qquad (8)$$

The last two terms on the right hand side of Equation (6) are the generation and disappearance of particles caused by breakage. The volume-based breakage kernel function $a(v)$ represents the breakage frequency of particles with volume v. Particle breakage is related to the particle size, velocity gradient, and particle concentration. The exponential breakage kernel function obtained by fitting the experimental data by Spicer [18] is:

$$a(v) = A\left(\frac{\varepsilon}{\nu}\right)^{0.8} v^{1/3} \qquad (9)$$

where A is 0.47 m^{-1}s$^{-0.6}$. In Equation (6), the particle breakage distribution $b(v \mid v_1)$ gives the relationship between the parent particles and separated sub-particles. Large particle breakage is not a simple process. The floc is assumed to be composed of small particles of the same size, meaning the distribution function of large, symmetrical broken particles composed of monomers is [19]:

$$b(v|v_1) = \begin{cases} 2 & v_1 = 2v \\ 0 & \text{else} \end{cases} \qquad (10)$$

2.3. Moment Equation of Particle Density

In order to obtain the number density distribution of particles more effectively, the general dynamic equation for nanoparticles is usually transformed into the moment equation of the particle density.

The k-th moment of the particle density is defined as:

$$m_k = \int_0^\infty n(v,t) v^k dv \qquad (11)$$

Based on Equation (11), Equation (6) can be transformed into a moment equation by multiplying the terms of Equation (6) by v^k and then integrating this over the entire volume distribution:

$$\begin{aligned}\frac{\partial m_k}{\partial t} + u_i \frac{\partial m_k}{\partial x_i} - \frac{\partial}{\partial x_i}\left[D_T \frac{\partial m_k}{\partial x_i}\right] &= \frac{1}{2}\int_0^\infty\int_0^\infty\left[(v+v_1)^k - v^k - v_1^k\right]\beta(v_1,v)n(v,t)n(v_1,t)dvdv_1 \\ &+ \int_0^\infty v^k \int_v^\infty a(v_1)b(v|v_1)n(v_1)dv_1 dv - \int_0^\infty v^k a(v) n(v) dv\end{aligned} \qquad (12)$$

Substituting Equations (7)–(11) into Equation (12), it can be found that different fractional moments in the equation are difficult to solve. Therefore, the Taylor series expansion technique [20] is used to transform the fractional moment into the moments represented by the first three moments (i.e., 0, 1, 2):

$$\frac{\partial m_0}{\partial t} + u_i \frac{\partial m_0}{\partial x_i} - \frac{\partial}{\partial x_i}\left(D_T \frac{\partial m_0}{\partial x_i}\right) = -\sqrt{\frac{3}{10\pi}\frac{\varepsilon}{\nu}} \frac{m_0^3 m_2^2 - 20 m_0^2 m_1^2 m_2 + 127 m_0 m_1^4}{27 m_1^3}$$
$$-0.47\left(\frac{\varepsilon}{\nu}\right)^{0.8} \frac{m_0^{5/3} m_2 - 10 m_0^{2/3} m_1^2}{9 m_1^{5/3}} + \frac{2kT}{3\mu}\left[\frac{2 m_0^4 m_2^2 - 13 m_0^3 m_1^2 m_2 - 151 m_0^2 m_1^4}{81 m_1^4}\right]$$
$$-\frac{1.591\lambda}{(3/4\pi)^{1/3}} \frac{-5 m_0^{13/3} m_2^2 + 64 m_0^{10/3} m_1^2 m_2 + 103 m_0^{10/3} m_1^4}{81 m_1^{13/3}} \tag{13}$$

$$\frac{\partial m_1}{\partial t} + u_i \frac{\partial m_1}{\partial x_i} - \frac{\partial}{\partial x_i}\left(D_T \frac{\partial m_1}{\partial x_i}\right) = 0 \tag{14}$$

$$\frac{\partial m_2}{\partial t} + u_i \frac{\partial m_2}{\partial x_i} - \frac{\partial}{\partial x_i}\left(D_T \frac{\partial m_2}{\partial x_i}\right) = -\sqrt{\frac{3}{10\pi}\frac{\varepsilon}{\nu}} \frac{4\left(5 m_0^2 m_2^2 + 14 m_1^4 + 35 m_0 m_1^2 m_2\right)}{27 m_0 m_1}$$
$$-0.47\left(\frac{\varepsilon}{\nu}\right)^{0.8} \frac{14 m_0 m_1^{1/3} m_2 - 5 m_1^{7/3}}{18 m_0^{4/3}} + \frac{4kT}{3\mu}\left[\frac{2 m_0^4 m_2^2 - 13 m_0 m_1^2 m_2 - 151 m_1^4}{81 m_1^2}\right]$$
$$+\frac{1.591\lambda}{(3/4\pi)^{1/3}} \frac{-4 m_0^{7/3} m_2^2 + 8 m_0^{4/3} m_1^2 m_2 + 320 m_0^{1/3} m_1^4}{81 m_1^{7/3}} \tag{15}$$

The zero-order moment m_0 represents the total number of particles of all sizes in the unit volume at a given position and time, which is also called the particle concentration. The first-order moment m_1 represents the volumes of all particles in the unit volume at a given position and time, which is also called the volume concentration. The second-order moment m_2 is related to the dispersion of particles. In the simulation, the dimensionless quantity is defined as $M_n = m_n/m_{nn}(n = 0, 1, 2)$, where m_{nn} is the initial value of m_n. The geometric mean diameter d_g of particles is defined as:

$$d_g = \left(\frac{m_1^2}{\sqrt{m_0^3 m_2}}\right)^{1/3} \tag{16}$$

3. Numerical Simulation

All simulations are based on the finite volume method and are carried out using the OpenFOAM platform. OpenFOAM is an open source CFD software and has an extensive range of features to solve anything from complex fluid flows involving turbulence and heat transfer, to solid mechanics and electromagnetic simulations. The equations of fluid flow are solved numerically with the basic solver pisoFoam in the platform. The piso algorithm takes an iterative approach to deal with the coupled pressure–velocity. The discrete and solution modes of the specific equation terms are also given. Gaussian linear interpolation scheme is employed to discretize divergence terms, gradient terms, and Laplace terms of the equations (interpolated from the body center of the grid unit to the surface center). The equations for nanoparticles are solved numerically with the self- made solver. The one-way coupling method is used, i.e., where the effect of particles on the flow is ignored. The particles are obtained from lit permethrin-based mosquito coils. The particles have a density of 730 kg/m^3 and a diameter of 98 nm, which are the typical density and diameter of nanoparticles.

3.1. Main Steps

The calculation domain is shown in Figure 1, where the boundary conditions are as follows. The inlet velocity is evenly distributed and equal to 2.664 m/s, the pressure boundary condition is adopted at the outlet, and the no slip boundary condition is adopted on the wall. In the simulation, the range of Reynolds numbers is $300 < Re < 3 \times 10^5$. In this Reynolds number range, the boundary layer on the cylinder surface is laminar but the flow behind the cylinder is turbulent. In addition, the flow around a cylinder with this range of Reynolds number is very common. In the simulation, the grid around a single cylinder is shown in Table 1, where 2D and 3D represent two and three dimension, respectively.

Since all boundaries are far enough from the wake region, the effects of the existence of boundaries (i.e., size of the domain) on the solution accuracy can be ignored.

Table 1. Mesh characteristics.

Case	Flow	Grid Number
A	2D	28,800
B	2D	46,000
C	3D	668,000

In the simulation, the time step ΔT is 1.5×10^{-4} s and the courant number is less than 1 for achieving numerical stability and accuracy. The tolerance set by the solver was 10^{-5}. The mean velocity and pressure are calculated by adding the field average functions to the control file.

3.2. Grid Independence Test and Validation

The velocity distribution of the flow around a cylinder is shown in Figure 2, where a stagnation point is formed at the front end of the cylinder. The region close to the tail of the cylinder is the wake region, where the velocity is very small. Downstream of cylinder, two alternating vortices obviously appear. In order to validate the numerical method and code, the present numerical result for the time-averaged streamline in 3D flow is compared with the experimental ones [9], as shown in Figure 3, where both results are in good agreement.

Figure 2. Velocity distribution of flow around a cylinder.

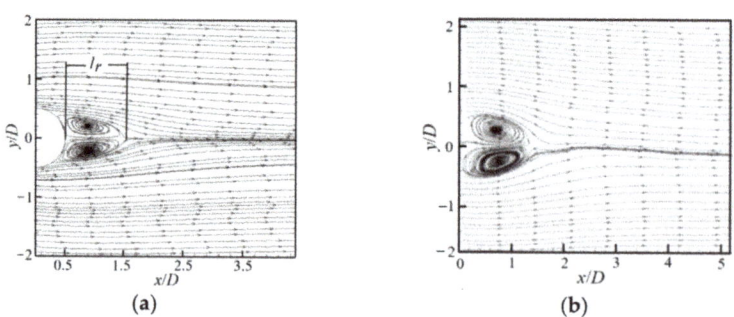

Figure 3. Distribution of time-averaged streamlines: (**a**) present result; (**b**) experimental result [9].

An important parameter to describe the flow around a cylinder is the reflux length l_r, which is defined as shown in Figure 3a. The present numerical result for the reflux length l_r is further compared with other numerical results. For the present result in 3D flow, $l_r = 0.08 - 0.025$ m $= 0.55$ m and the particle diameter D = 0.5 m, so $l_r = 1.1D$ ($Re = 9000$), as shown in Figure 4. Keita et al. [8] also numerically simulated the flow around a 2D cylinder at $Re = 9300$ and gave a reflux length $l_r = 0.9D$. The main reason for the slight deviation in the value of the reflux length is the difference between the Reynolds number and calculation dimension. Figure 4 shows a comparison of the streamwise velocity at the centerline along the x direction, where it can be seen that the present numerical results are in good agreement with the experimental results [21].

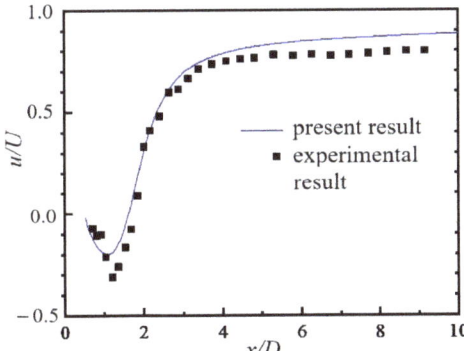

Figure 4. Comparison of streamwise velocity at the centerline.

A grid independence test is performed by changing the grid number, as shown in Figure 5, where mesh A and mesh B correspond to 28,800 and 46,000 grid numbers, respectively, as shown in Table 1. The results are almost the same for both grid numbers, so mesh A is selected in the simulation.

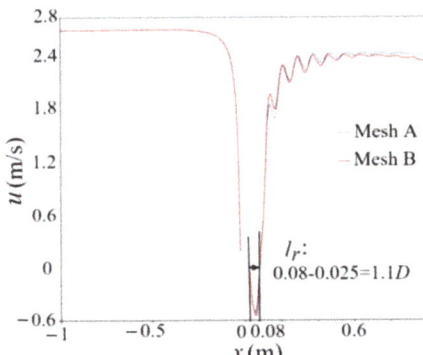

Figure 5. Streamwise velocity at the centerline along the x direction.

4. Results and Discussion

4.1. The Flow around a Single Cylinder

4.1.1. Particle Coagulation and Distribution of Particle Concentration

In a fully developed turbulent wake flow, the particle coagulation mainly results from the Brownian motion and turbulent shear in the free molecular region. The coagulated particles may break up under the action of turbulent shear. The characteristic times for flow convection, particle coagulation, and particle breakage are different. Previous research results have shown that particle coagulation occurs when the ratio of the characteristic time of flow convection to that of particle coagulation is less than 0.1 [22–24]. Figure 6 shows the distributions of particle concentration M_0 ($=m_0/m_{00}$) with different initial particle concentrations m_{00}. It can be seen that the values of M_0 upstream of the cylinder are uniformly distributed along the spanwise direction and gradually reduced along the flow direction, which indicates that coagulation has occurred in the process of particle transportation downstream. The values of M_0 are large (as shown in Figure 6a) and small (as shown in Figure 6b) in the wake region close to the tail of the cylinder, indicating that it is easy for the particles with an initial $m_{00} = 3.6 \times 10^{11}/\text{m}^3$ (i.e., relatively low initial particle concentration) to enter the wake area behind the cylinder, although the opposite is true for the particles with an initial $m_{00} = 3.6 \times 10^{15}/\text{m}^3$. The variation ranges of M_0 are 0.9997~1.0002, as shown in Figure 6a, and 0.23~1.0, as shown in Figure 6b, showing that it is easier for the particles to coagulate when m_{00} is high, resulting in a wide distribution range of M_0. An oscillating wake is formed behind the cylinder and the values of M_0 in the oscillating wake are obviously larger (as shown in Figure 6a) and smaller (as shown in Figure 6b) than that around the wake. At the position close to the outlet, the values of M_0 return to the uniform distribution along the spanwise direction. Three conclusions can be drawn from Figure 6: (1) there is an obvious coagulation phenomenon of particles for the parameters of m_{00}, d_0, and Re given in the paper, and the larger the value of m_{00}, the larger the differences in values of M_0 in different regions of the flow; (2) the existence of the cylinder has a great influence on the distribution of the particles; (3) m_{00} will affect the number of particles in the cylindrical wake.

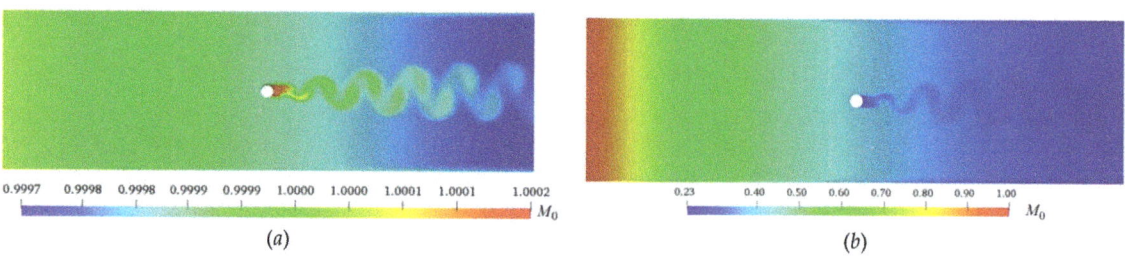

Figure 6. Distribution of M_0 values (d_0 = 98 nm, Re = 9000): (**a**) $m_{00} = 3.6 \times 10^{11}/\text{m}^3$; (**b**) $m_{00} = 3.6 \times 10^{15}/\text{m}^3$.

4.1.2. Function of Particle Breakage

The coagulated particles may break up under the action of turbulent shear. However, whether the coagulated particles are broken depends on the size of coagulated particles and the shear rate of the flow. The variations in particle concentration M_0 ($=m_0/m_{00}$) at the centerline along the flow direction are shown in Figure 7. In the flow upstream of the cylinder ($x < 0$), the values of M_0 are decreased along the flow direction due to the occurrence of particle coagulation, and there is almost no difference in M_0 between the cases with and without considering particle breakage because the particles do not have enough time to break up after coagulation and the shear rate of the flow is very small. The discontinuous part of M_0 is located at the position of the cylinder ($x = 0$). In the flow downstream of the cylinder, the values of M_0 for the case considering particle breakage are obviously larger than that for the case without considering particle breakage because the shear rate of the flow downstream of the cylinder is large. A large shear rate is more likely to lead to particle breakage and an increase in M_0. The fluctuation curve shows that the values of M_0 are affected by the wake flow. Therefore, putting an obstacle in a uniform flow will promote particle breakage under a certain particle concentration.

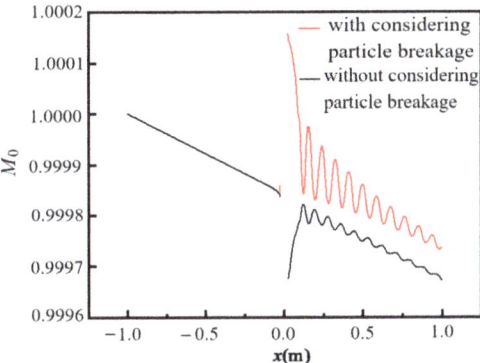

Figure 7. Variations in M_0 at the centerline along the flow direction (Re = 9000, d_0 = 98 nm, m_{00} = 3.6 × 10^{11}/m³).

4.1.3. Distribution of Particles along the Spanwise Direction

Figure 8 shows the distribution of M_0 along the spanwise direction in the flow downstream of the cylinder for two different m_{00}. In Figure 8a, the values of M_0 for both cases with and without considering particle breakage decrease along the flow direction, showing that the particle coagulation effect is larger than the breakage effect. The values of M_0 along the spanwise direction, except for the wake regions behind the cylinder and the near wall region, are uniformly distributed and the same for the cases with and without considering particle breakage because the shear rate of the flow is very small in this region.

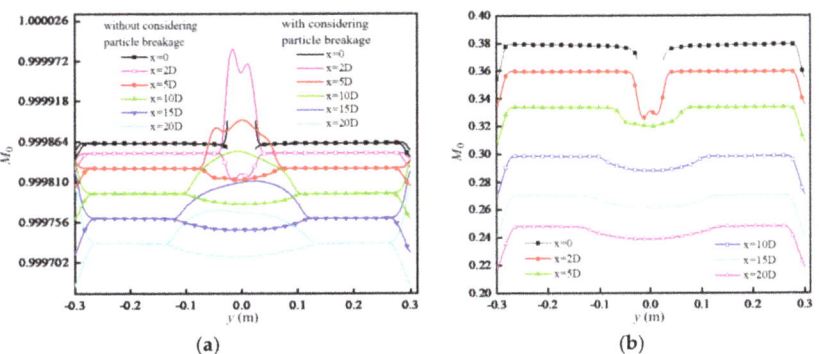

Figure 8. Distribution of M_0: (**a**) m_{00} = 3.6 × 10^{11}/m³; (**b**) m_{00} = 3.6 × 10^{15}/m³ (considering particle breakage).

In the wake region behind the cylinder, the values of M_0 for the case considering particle breakage are obviously larger than that for the case without considering particle breakage because the shear rate in this region is large. As the flow develops downstream, the values of M_0 for both cases with and without considering particle breakage tend to be uniformly distributed along the spanwise direction. Figure 8b shows the distribution of M_0 when considering particle breakage in the case of higher m_{00}. It can be seen that the distribution of M_0 is qualitatively consistent with that in Figure 8a, but the magnitude of M_0 is far less than that in Figure 8a. The reason is that the higher m_{00} increases the chance of particle coagulation, so the value of M_0 (=m_0/m_{00}) is far less than m_{00}.

4.2. The Flow around Multiple Cylinders

4.2.1. Particle Coagulation and Distribution of Particles

For the flow around multiple cylinders, the distribution of M_0 is shown in Figure 9. It can be seen that the values of M_0 upstream of the cylinders are uniformly distributed along the spanwise direction and gradually reduced along the flow direction because of particle coagulation. The values of M_0 behind the cylinders oscillate laterally under the influence of the flow structure in the wake of each cylinder. Compared with the flow around a single cylinder, as shown in Figure 5, the wake region of the flow around multiple cylinders is shorter along the flow direction due to the mutual interference of wakes behind multiple cylinders. The distribution of M_0 along the spanwise direction becomes uniform as the influence of the wake disappears, and turns to a parabolic distribution due to the increase in wall influence at the position close to the outlet.

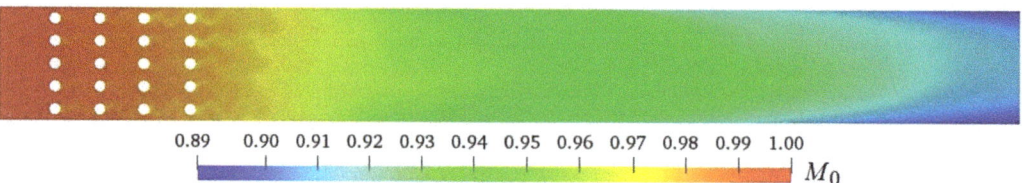

Figure 9. Distribution of M_0 in the flow around multiple cylinders (d_0 = 98 nm, m_{00} = 3.6 × 10^{14}/m^3, Re = 9000).

4.2.2. Distribution of the Geometric Mean Diameter of Particles with the Initial Monodispersity

Figure 10 shows the distribution of the geometric mean diameters d_g of particles along the flow and spanwise directions for initial monodisperse particles. The values of d_g are given along the centerline of the flow in Figure 10a, where there are four jumps and discontinuities in the values of d_g at the positions of the four columns of cylinders. The values of d_g increase along the flow direction because of the occurrence of particle coagulation in the process of particle transportation downstream, showing that the characteristic time of the flow convection is longer than the characteristic time of the particle coagulation and that the particles have enough time to coagulate, while at the same time showing that the effect of particle coagulation is larger than that of particle breakage. The values of d_g are larger for the case without considering particle breakage than that when considering particle breakage, which is reasonable because the number of small particles increases when considering particle breakage. Figure 10b shows the distribution of d_g along the spanwise direction at different positions of x. The values of d_g increase along the flow direction and fluctuate along the spanwise direction. The closer to the cylinders, the more frequent the fluctuations of d_g due to the influence of the wake. In the far downstream area (x = 0.90), the distribution of d_g along the spanwise direction shows a single arc due to the influence of the boundary. The values of d_g are small in the middle and large on both sides at x = 0.50 and x = 0.90, which is the reason that the breakage of coagulated particles is larger in the middle area with high M_0.

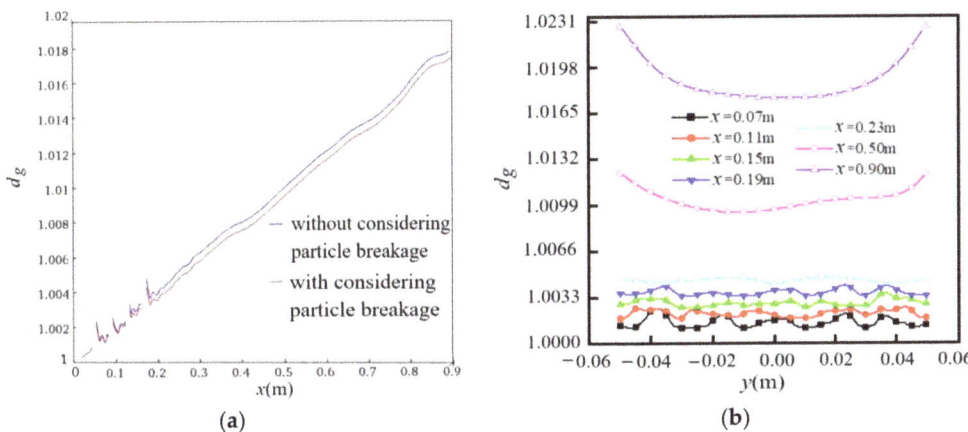

Figure 10. Distribution of d_g for initial monodisperse particles (d_0 = 98 nm, m_{00} = 3.6×10^{14}/m^3, Re = 9000): (**a**) along the flow direction; (**b**) along the spanwise direction (considering particle breakage).

4.2.3. Distribution of Geometric Mean Diameter of Particles with Initial Polydispersity

Figure 11 shows the distribution of the geometric mean diameter d_g of particles along the flow and spanwise directions for initial polydisperse particles with d_0 = 98 nm and geometric standard deviation σ = 1.65. In Figure 11a, the variations of d_g along the flow direction show the same trend as in Figure 10a, although the difference is that the values of d_g are almost the same for the cases with and without considering particle breakage for the initial polydisperse particles. This is because the particle breakage distribution $b(v|v_1)$ included in Equation (6) is only for the parent particles composed of separated sub-particles, with the same size as that shown in Equation (10). The particle breakage is insignificant for the polydisperse particles of different sizes. In Figure 11b, the distribution of d_g along the spanwise direction at different positions of x is similar to that shown in Figure 10b; the difference is that the distribution of d_g along the spanwise direction is flatter in Figure 11b than in Figure 10b.

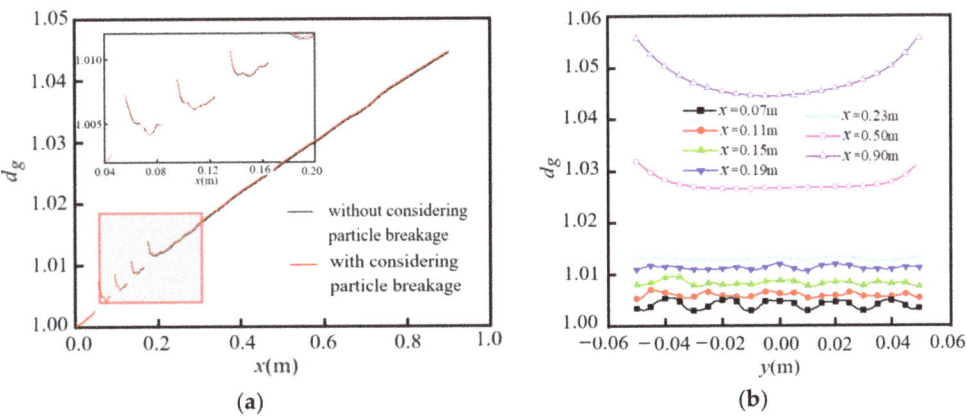

Figure 11. Distribution of d_g for initial polydisperse particles (d_0 = 98 nm, m_{00} = 3.6 × 10^{14}/m^3, Re = 9000): (**a**) along the flow direction; (**b**) along the spanwise direction (considering particle breakage).

5. Conclusions

In this paper, the Reynolds averaged N-S equation and general dynamic equation for nanoparticles are numerically solved in the two-phase flow around a single cylinder and multiple cylinders. The distributions of M_0 and d_g values of particles with different m_{00} and d_0 values are given. Some of the results are validated by comparing them with the experimental and numerical results. The effects of particle coagulation and breakage, m_{00} and d_0, on the particle distribution are discussed. The main conclusions are summarized as follows:

(1) For the flow around a single cylinder, there is an obvious particle coagulation phenomenon. The existence of a single cylinder has a great influence on the distribution of particles. The number of particles in the wake is dependent on the value of m_{00}. In the flow upstream of the cylinder, there is almost no difference in M_0 between the cases with and without considering particle breakage. Putting a cylinder in a uniform flow will promote particle breakage. As the flow develops downstream, the values of M_0 for both cases with and without considering particle breakage tend to be uniformly distributed along the spanwise direction;

(2) For the flow around multiple cylinders, the values of M_0 are reduced along the flow direction upstream of the cylinders and oscillate laterally behind the cylinders under the influence of the flow structure. For the initial monodisperse particles, the effect of particle coagulation is larger than that of particle breakage. For the initial polydisperse particles with $d_0 = 98$ nm and geometric standard deviation $\sigma = 1.65$, the variations in d_g show the same trend as for the initial monodisperse particles, but the differences are that the values of d_g are almost the same for the cases with and without considering particle breakage;

(3) In future work, it will be necessary to further study the numerical simulation of three-dimensional flow and to explore the particle breakage model in the case of polydisperse particles.

Author Contributions: Conceptualization, J.L. and R.S.; methodology, R.S. and H.Y.; software, R.S. and H.Y.; validation, R.S. and H.Y.; writing, R.S. and H.Y.; resources R.S. and J.L.; review, J.L. All authors have read and agreed to the published version of the manuscript.

Funding: This work was supported by the National Natural Science Foundation of China (Grant no. 12132015).

Institutional Review Board Statement: Not applicable.

Informed Consent Statement: Not applicable.

Conflicts of Interest: There are no conflicts of interest regarding the publication of this paper.

References

1. Zhou, H.; Mo, G.; Cen, K. Numerical investigation of dispersed gas-solid two-phase flow around a circular cylinder using lattice Boltzmann method. *Comput. Fluids* **2011**, *52*, 130–138. [CrossRef]
2. Haddadi, H.; Shojaei-zadeh, S.; Morris, J.F. Lattice-Boltzmann simulation of inertial particle-laden flow around an obstacle. *Phys. Rev. Fluids* **2016**, *1*, 024201. [CrossRef]
3. Jafari, S.; Salmanzadeh, M.; Rahnama, M.; Ahmadi, G. Investigation of particle dispersion and deposition in a channel with a square cylinder obstruction using the lattice Boltzmann method. *J. Aerosol Sci.* **2010**, *41*, 198–206. [CrossRef]
4. Haddadi, H.; Shojaei-Zadeh, S.; Connington, K.; Morris, J.F. Suspension flow past a cylinder: Particle interactions with recirculating wakes. *J. Fluid Mech.* **2014**, *760*, R2. [CrossRef]
5. Huang, Q.; Zhang, Y.; Yao, Q.; Li, S.Q. Numerical and experimental study on the deposition offine particulate matter during the combustion of pulverized lignite coal in a 25 kW combustor. *Powder Technol.* **2017**, *317*, 449–457. [CrossRef]
6. Jeong, S.; Kim, D. Simulation of the particle deposition on a circular cylinder in high-temperature particle-laden flow. *Korean Soc. Manuf. Process Eng.* **2019**, *18*, 73–81. [CrossRef]
7. Gopan, A.; Yang, Z.; Axelbaum, R.L. Predicting particle deposition for flow over a circular cylinder in combustion environments. *Proc. Combust. Inst.* **2019**, *37*, 4427–4434. [CrossRef]
8. Keita, N.S.; Mehel, A.; Murzyn, F.; Taniere, A.; Arcen, B.; Diourte, B. Numerical study of ultrafine particles dispersion in the wake of a cylinder. *Atmos. Pollut. Res.* **2019**, *10*, 294–302. [CrossRef]

9. Tu, C.; Zhang, J. Nanoparticle-laden gas flow around a circular cylinder at high Reynolds number. *Int. J. Numer. Methods Heat Fluid Flow* **2014**, *24*, 1782–1794. [CrossRef]
10. Liu, H.; Yang, F.; Tan, H.; Li, Z.H.; Feng, P.; Du, Y.L. Experimental and numerical investigation on the structure characteristics of vortex generators affecting particle agglomeration. *Powder Technol.* **2020**, *362*, 805–816. [CrossRef]
11. Kolsi, L.; Selimefendigil, F.; Oztop, H.F.; Hassen, W.; Aich, W. Impacts of double rotating cylinders on the forced convection of hybrid nanofluid in a bifurcating channel with partly porous layers. *Case Stud. Therm. Eng.* **2021**, *26*, 101020. [CrossRef]
12. Alsabery, A.I.; Selimefendigil, F.; Hashim, I.; Chamkha, A.J.; Ghalambaz, M. Fluid-structure interaction analysis of entropy generation and mixed convection inside a cavity with flexible right wall and heated rotating cylinder. *Int. J. Heat Mass Transf.* **2019**, *140*, 331–345. [CrossRef]
13. Ramkrishna, D. *Population Balances: Theory and Applications to Particulate Systems in Engineering*; Academic Press: London, UK, 2000.
14. Vasudev, V.; Ku, X.K.; Lin, J.L. Kinetic study and pyrolysis characteristics of algal and lignocellulosic biomasses. *Bioresour. Technol.* **2019**, *288*, 121496. [CrossRef] [PubMed]
15. Guichard, R.; Belut, E. Simulation of airborne nanoparticles transport, deposition and aggregation: Experimental validation of a CFD-QMOM approach. *J. Aerosol Sci.* **2017**, *104*, 16–31. [CrossRef]
16. Yu, M.; Lin, J.Z. Taylor-expansion moment method for agglomerate coagulation due to Brownian motion in the entire size regime. *J. Aerosol Sci.* **2009**, *40*, 549–562. [CrossRef]
17. Saffman, P.G.; Turner, J.S. On the collision of drops in turbulent clouds. *J. Fluid Mech.* **1956**, *1*, 16–30. [CrossRef]
18. Spicer, P.T.; Pratsinis, S.E. Coagulation and fragmentation: Universal steady-state particle-size distribution. *AIChE J.* **1996**, *42*, 1612–1620. [CrossRef]
19. Lick, W.; Lick, J. Aggregation and disaggregation of fine-grained lake sediments. *J. Great Lakes Res.* **1988**, *14*, 514–523. [CrossRef]
20. Yu, M.; Lin, J.; Chan, T. A new moment method for solving the coagulation equation for particles in Brownian motion. *Aerosol Sci. Technol.* **2008**, *42*, 705–713. [CrossRef]
21. Lam, K.; Wang, F.H.; So, R.M.C. Three-dimensional nature of vortices in the near wake of a wavy cylinder. *J. Fluids Struct.* **2004**, *19*, 815–833. [CrossRef]
22. Lin, J.Z.; Qian, L.J.; Xiong, H.B. Effects of operating conditions on the droplet deposition onto surface in the atomization impinging spray with an impinging plate. *Surf. Coat. Technol.* **2009**, *203*, 1733–1740. [CrossRef]
23. Qian, L.J.; Lin, J.Z.; Xiong, H.B. A fitting formula for predicting droplet mean diameter for various liquid in effervescent atomization spray. *J. Therm. Spray Technol.* **2010**, *19*, 586–601. [CrossRef]
24. Yu, M.Z.; Lin, J.Z. Binary homogeneous nucleation and growth of water-sulfuric acid nanoparticles using a TEMOM model. *Int. J. Heat Mass Transf.* **2010**, *53*, 635–644. [CrossRef]

MDPI
St. Alban-Anlage 66
4052 Basel
Switzerland
Tel. +41 61 683 77 34
Fax +41 61 302 89 18
www.mdpi.com

Entropy Editorial Office
E-mail: entropy@mdpi.com
www.mdpi.com/journal/entropy

www.ingramcontent.com/pod-product-compliance
Lightning Source LLC
LaVergne TN
LVHW070641100526
838202LV00013B/856